The Complete Idiot's Reference Card

	1	2		3	4	5	6	7	8	9	10	11	12	13	14	15	16	17	18
1	1 H 1.01																		2 He 4.00
2	3 Li 6.94	4 Be 9.01												5 B 10.81	6 C 12.01	7 N 14.01	8 O 16.00	9 F 19.00	10 Ne 20.18
3	11 Na 22.99	12 Mg 24.31												13 Al 26.98	14 Si 28.09	15 P 30.97	16 S 32.07	17 Cl 35.45	18 Ar 39.95
4	19 K 39.10	20 Ca 40.08		21 Sc 44.96	22 Ti 47.87	23 V 50.94	24 Cr 52.00	25 Mn 54.94	26 Fe 55.85	27 Co 58.93	28 Ni 58.69	29 Cu 63.55	30 Zn 65.39	31 Ga 67.72	32 Ge 72.61	33 As 74.92	34 Se 78.96	35 Br 79.90	36 Kr 83.80
5	37 Rb 85.47	38 Sr 87.62		39 Y 88.91	40 Zr 91.22	41 Nb 92.91	42 Mo 95.94	43 Tc (98)	44 Ru 101.07	45 Rh 102.91	46 Pd 106.42	47 Ag 107.87	48 Cd 112.41	49 In 114.82	50 Sn 118.71	51 Sb 121.76	52 Te 127.60	53 I 126.90	54 Xe 131.29
6	55 Cs 132.91	56 Ba 137.33	57-70*	71 Lu 174.97	72 Hf 178.49	73 Ta 180.95	74 W 183.84	75 Re 186.21	76 Os 190.23	77 Ir 192.22	78 Pt 195.08	79 Au 196.97	80 Hg 200.59	81 Tl 204.38	82 Pb 207.2	83 Bi 208.98	84 Po (209)	85 At (210)	86 Rn (222)
7	87 Fr (223)	88 Ra (226)	89-102**	103 Lr (262)	104 Rf (261)	105 Db (262)	106 Sg (266)	107 Bh (264)	108 Hs (277)	109 Mt (268)	110 Ds (271)	111 Rg (272)							

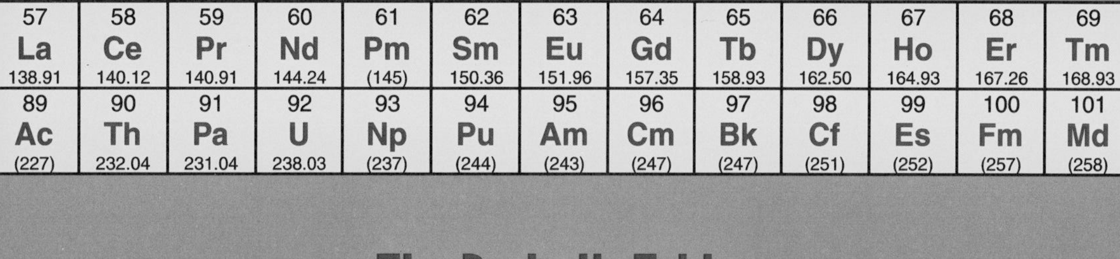

*	57 La 138.91	58 Ce 140.12	59 Pr 140.91	60 Nd 144.24	61 Pm (145)	62 Sm 150.36	63 Eu 151.96	64 Gd 157.35	65 Tb 158.93	66 Dy 162.50	67 Ho 164.93	68 Er 167.26	69 Tm 168.93	70 Yb 173.04
**	89 Ac (227)	90 Th 232.04	91 Pa 231.04	92 U 238.03	93 Np (237)	94 Pu (244)	95 Am (243)	96 Cm (247)	97 Bk (247)	98 Cf (251)	99 Es (252)	100 Fm (257)	101 Md (258)	102 No (259)

The Periodic Table

ALPH

Elements in the same family in the periodic table have similar properties. Some of the most important families are these:

- Group 1 (except for hydrogen)—Alkali Metals: Alkali metals are highly reactive, combining readily with air and water. Though they are metallic, their densities are low (only rubidium and cesium are denser than water) and are soft enough to be cut with a knife. The high reactivity of the alkali metals comes from the fact that they have only one more electron than the very stable noble gases. As a result, they react vigorously in attempts to lose this extra electron. Alkali metals can be found in sodium vapor fog lamps and in the psychiatric drug lithium carbonate.
- Group 2—Alkaline Earth Metals: The alkaline earth metals have many of the same properties as the alkali metals, although they are less extreme. For example, most alkaline earth metals react with air and water, but much less violently than the alkali metals. Alkaline earth metals are generally harder than the alkali metals, but are still softer than many other metals. The diminished reactivity of the alkaline earth metals can also be explained by their electron configurations. Because they have to lose two electrons to become like a noble gas, they are somewhat less reactive than the alkali metals.
- Groups 3–12—d-Transition Metals (frequently called simply "transition metals"): Though properties of the d-transition elements vary greatly, many of them are hard, have high melting and boiling points, are excellent conductors of heat and electricity, and have moderate to low reactivities. Transition metals are used for a variety of purposes such as structural materials in buildings, power transmission lines, jewelry, and knives.
- Group 17—Halogens: These are highly reactive elements that combine readily with metals to form salts. This extremely high reactivity comes from their electron configurations—because they need only one more electron to have the electron configurations of a noble gas, they react vigorously to pick up that electron whenever possible. The halogens are diatomic elements, meaning that they have the general formula X_2 (for example, fluorine exists as F_2 in its pure form). Fluorine and chlorine are gases under standard conditions, while bromine is a liquid and iodine is a solid. Halogens are widely used in water treatment, in the manufacturing of other chemicals, and in plastics such as Teflon.
- Group 18—Noble gases: Noble gases are almost entirely unreactive. This lack of reactivity stems from the fact that completely filled s- and p-orbitals (see Chapter 5) make them very stable. As a result, very few noble gas compounds can be made. Noble gases are used in advertising signs, toy balloons and blimps, and as inert atmospheres in locations where chemical reactions would be undesirable.
- f-Transition metals (sometimes called "inner transition metals"): The f-transition elements consist of the two rows at the bottom of the periodic table, and aren't properly said to be in any of the 18 "groups." The top row, also known as the lanthanides, consists of shiny, reactive metals. Because many lanthanides emit colored light when hit by a beam of electrons, they are used as phosphors in television sets and fluorescent light bulbs. The bottom row, also known as actinides, are primarily radioactive elements that have a wide variety of uses such as nuclear fuel sources, smoke detectors, and atomic bombs.
- Hydrogen—The weirdo: Hydrogen has properties unlike any other element in the periodic table. Though it's found in the metallic region of the periodic table, it is a nonmetallic gas. It is diatomic, found as H_2. Hydrogen reacts slowly with other elements at room temperature but may react blindingly fast when heated or catalyzed. Hydrogen is used in the manufacture of ammonia, sulfuric acid, and methanol, and is widely discussed as a fuel alternative to gasoline.

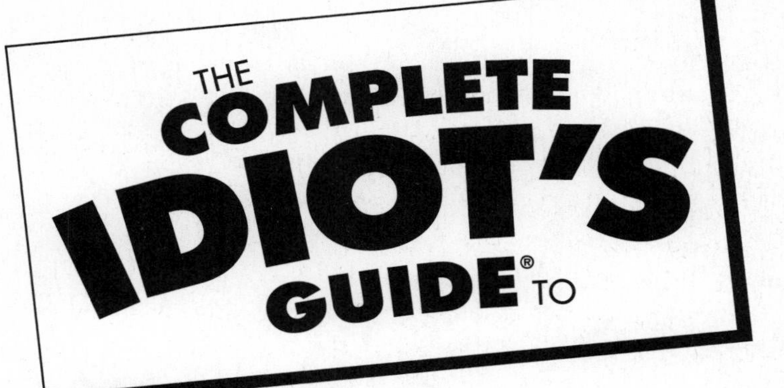

Chemistry

Second Edition

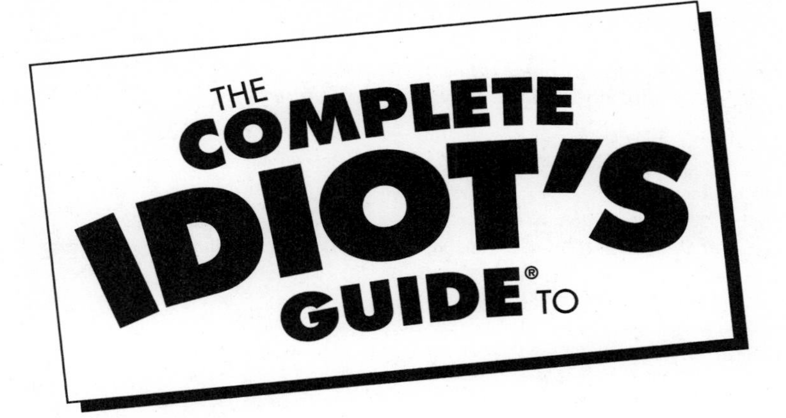

Chemistry

Second Edition

by Ian Guch

A member of Penguin Group (USA) Inc.

To Ingrid for her unconditional support, my friends for watching out for me, and my students for keeping me on my toes.

ALPHA BOOKS

Published by the Penguin Group

Penguin Group (USA) Inc., 375 Hudson Street, New York, New York 10014, U.S.A.

Penguin Group (Canada), 10 Alcorn Avenue, Toronto, Ontario, Canada M4V 3B2 (a division of Pearson Penguin Canada Inc.)

Penguin Books Ltd, 80 Strand, London WC2R 0RL, England

Penguin Ireland, 25 St Stephen's Green, Dublin 2, Ireland (a division of Penguin Books Ltd)

Penguin Group (Australia), 250 Camberwell Road, Camberwell, Victoria 3124, Australia (a division of Pearson Australia Group Pty Ltd)

Penguin Books India Pvt Ltd, 11 Community Centre, Panchsheel Park, New Delhi—110 017, India

Penguin Group (NZ), cnr Airborne and Rosedale Roads, Albany, Auckland 1310, New Zealand (a division of Pearson New Zealand Ltd)

Penguin Books (South Africa) (Pty) Ltd, 24 Sturdee Avenue, Rosebank, Johannesburg 2196, South Africa

Penguin Books Ltd, Registered Offices: 80 Strand, London WC2R 0RL, England

International Standard Book Number: 1-59257-514-5
Library of Congress Catalog Card Number: 2006920726

08 8 7 6

Interpretation of the printing code: The rightmost number of the first series of numbers is the year of the book's printing; the rightmost number of the second series of numbers is the number of the book's printing. For example, a printing code of 06-1 shows that the first printing occurred in 2006.

Printed in the United States of America

Publisher: *Marie Butler-Knight*
Editorial Director: *Mike Sanders*
Managing Editor: *Billy Fields*
Development Editor: *Michael Thomas*
Senior Production Editor: *Janette Lynn*
Copy Editor: *Sara Bosin*
Illustrator: *Chris Eliopoulos*
Cover Designer: *Bill Thomas*
Book Designers: *Trina Wurst and Kurt Owens*
Indexer: *Brad Herriman*
Layout: *Becky Harmon*
Proofreader: *Mary Hunt*

Contents at a Glance

Contents

Appendixes

Foreword

Finally, a chemistry book for "chemophobes"! Thanks to *The Complete Idiot's Guide* people, you can now have fun with chemistry. And it's safe fun, too—no test tubes or pH meters. All you need is the book you're holding in your hands.

Of all the subjects I took in high school and college, chemistry has provided the most vivid memories for me. In high school, there was this guy who always showed up to chemistry class covered in fish guts and stinking to high heaven. The teacher was not amused and repeatedly kicked him out of class. That stinky fisherman was my best friend Bill, who I admired so much that I subsequently joined his band, the Descendents. Then there was the student who had obviously been paying attention during the alkali metals lecture, and tossed the entire lab supply of elemental sodium into the fish pond outside the classroom … BOOM! The pond erupted as the sodium reacted with water, and several guppies and carp were temporarily … displaced. (Don't try this at home, kids.)

In college, I sweated over organic chemistry and ended up with a 100 percent on the first exam. I can still remember the warm glow I felt as students around me glared and muttered, "Thanks for destroying the curve, jerk!" Another bizarre test result I remember was the 45 percent score that actually counted as an "A" when the class curve was calculated. When I expressed amazement at the dismal overall curve to my more jaded classmates, they replied, "What'd you expect, it's O-Chem …"

Despite my success in chemistry class, I never really caught the chemistry bug. If only this book had been around when I was in school! This is a chemistry book designed to inspire. It achieves this through humor, examples of chemistry in everyday life, and enthusiasm for the subject that only a top-notch chemistry instructor can provide. No doubt many readers of this book have approached chemistry warily; by the time they are finished they may have indeed caught the chemistry bug and will be signing up for Chem 1A. Or if you're one of those people who break out in a cold sweat at the mere mention of chemistry, this book should go a long way toward curing your "chemophobia."

As with the entire *Complete Idiot's Guide* series, the book is written at a level that even a complete novice to chemistry can understand. Mr. Guch brings the reader gently into each chapter, often with a commonplace anecdote that allows an instant connection with the material. Especially helpful are the example problems the reader is asked to solve, with answers in the back. In chemistry, as with many other scientific disciplines, solving example problems is critical to the student's understanding of key concepts. Also included with the book is a glossary of chemical terms, which should calm the nerves of those students who treat chemistry as a "foreign language requirement."

What makes this book especially approachable is Mr. Guch's off-the-wall sense of humor, as in the chapter where he divulges his secret chili recipe to introduce the subject of chemical equations. (Let's be honest, we all know what chemical reactions your chili really causes, Mr. Chemistry Professor!)

As a supplement to your typical chemistry textbook, you can't go wrong with this book. It makes learning chemistry basics fun, and perhaps will inspire you to dust off that old chemistry tome and learn in depth about certain subjects.

And even if you never plan on extracting a sample, determining a pH, or titrating for a profession, this book can help get you through the hurdle of chemistry class. If you're a student with a problem, and that problem is a dreary, complex chemistry textbook, *The Complete Idiot's Guide to Chemistry, Second Edition*, may provide a "solution."

Enjoy!

Milo J. Aukerman, Ph.D.
Research biologist, E. I. du Pont de Nemours and Company
(and punk rock singer)

Introduction

Before you get started reading this book, I've got to tell you that I'm on to your little secret: you don't like chemistry. You probably bought this book because your chemistry teacher is boring, or because your textbook is too hard to understand, or because you're lousy with math and feel like you'll never pass without divine intervention. Unfortunately, since divine intervention doesn't usually work for things like this, you're stuck buying a book.

Relax. I understand what you're feeling. When I took my first chemistry course in high school, I felt that it was the most confusing class I'd ever taken, and that only a genius could understand what "moles" and "stoichiometry" were. It didn't help that my teacher was a crazy dude who wore a lime green lab coat and tended to rant. I swore I was done with chemistry forever.

In other words, you're in good hands. I know what it's like to be confused by chemistry, and I've got a bunch of tricks up my metaphorical sleeve to help you get through this with a minimum of hair pulling and screaming. That's fortunate, as I've found that many people find both hair pulling and screaming to be unnerving.

This doesn't mean that chemistry is the easiest class you'll ever take and that you're a moron for not understanding it sooner. Chemistry is challenging, and there's an awful lot you need to understand before it starts to really come together. However, we're going to take it in small chunks so your brain will have time to recover before moving on to the next topic. Trust me, I don't want this to be any worse than it has to be.

By the end, it's my greatest hope that you'll love chemistry as much as I do, decide to devote your life to it, and become a Nobel Prize–winning chemist. However, I have a gut feeling that many more of you will probably be happy to survive chemistry with a reasonable grade. I can't promise the Prize, but I can guarantee you that a reasonable grade is well within your grasp.

How This Book Is Organized

I've conveniently broken this book into six sections:

Part 1, "Firing Up the Bunsen Burner," introduces you to some of the basic concepts in chemistry. You'll learn about the history of chemistry and get to know the funny-looking guys who made it the science it is today. You'll learn about atoms, unit conversions, and the scientific method. In short, the first section is a delightful mix of introductory concepts that I think you'll understand without too much screaming.

Part 2, "A Matter of Organization," introduces you to concepts such as the periodic table, the types of matter, ionic and covalent compounds, and the dreaded mole. Though you may wonder why you should care about these topics, we'll spend some time examining how they actually make the life of a chemist easier. Believe it or not, you're a chemist now!

We get down to the real meat of chemistry in **Part 3, "Solids, Liquids, and Gases."** You may already think you know everything there is to understand about the three phases of matter. However, you'll quickly learn that, much like a demolition derby or punk rock show, there are subtleties that make what initially sounds incomprehensible into a wonderful and rewarding experience.

You're probably thinking to yourself that we haven't yet mentioned anything about chemical reactions. Well, we're not going to discuss the finer points of chemical reactions in **Part 4, "How Reactions Occur,"** but we will learn about some things that all chemical reactions have in common with each other, such as kinetics, stoichiometry, and equilibria. Not only that, but you'll learn to balance equations like a pro—if there were professional equation balancers, that is.

In **Part 5, "Chemical Reactions,"** we'll finally learn about each of the main types of chemical reaction that you're likely to need to know about. We'll take a look at topics as diverse as nuclear weapons, buffers, and organic chemistry. As the saying goes, these are a few of my favorite things.

Finally, as a delightful dessert to this chemical meal, we've got **Part 6, "Thermodynamics 101."** If you've ever wondered about things such as entropy, enthalpy, and free energy, this section is for you. If you've never wondered about these things, well, you're stuck reading about them anyway, because they're an important part of chemistry. I think you'll find that they're a lot easier than they sound—most big words are.

Helpful Reminders and Random Thoughts

Occasionally, a random or useful thought pops into my head and starts screaming to come out. Because the screaming in my head distracts me from writing, I've included these as sidebars in the main text of this book:

The Mole Says

These boxes highlight the most important things you'll need to know in the chapter.

def•i•ni•tion

These boxes provide definitions of difficult or unfamiliar words.

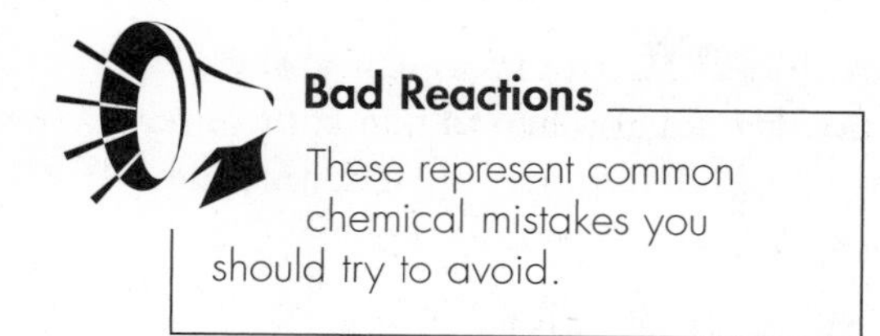

Bad Reactions

These represent common chemical mistakes you should try to avoid.

You've Got Problems

These are practice problems to help you better understand the material covered in each section.

Chemistrivia

These boxes provide interesting, bet-you-didn't-know tidbits of trivia.

Acknowledgments

Most books have a section in the front where the author mentions all of the people that used their intelligence and huge quantities of work to make the book into something special. Though I'd like to say that I did everything without any help at all from anyone, the fact of the matter is that there was a veritable army of people who turned my lousy rough copy into the polished masterpiece you're reading today.

As of five months ago, I had no idea that I was going to write a book. I'd like to thank Jessica Faust from BookEnds for tracking me down and convincing me that I was the right guy for the job. I'd also like to thank the folks at Alpha who gave the book the thumbs up and spent immeasurable time working on making it wonderful. Special thanks to Michael Thomas, who answered my thousands of questions and organized the army of people who made this book the work of art it is today.

I'd also like to thank the vast army of people who edited my rough draft into something I could send to the publisher. Special thanks go to my wife Ingrid (who was stuck reading the first draft), and my editing crew Sarah Aldemeyer, Ashley Bevis, Allix Bradecamp, Jillian Brems, Emily Derr, Killeen Hansen, Meaghan Hollenbeck, Sarah Holtz, Katie MacTurk, Mary Murphy, Jenny Olszewski, Beth Seifert, Cayley Sheehan, Kristina Slekys, Christine Tadie, and Tatjana Wiese (who were stuck editing the second draft). Without their help, this book would be a gigantic mass of run-on sentences and mangled punctuation.

More thanks to Bruce Rickborn, who edited the book for technical accuracy. I think it's safe to say that he not only made this book into a paragon of chemical accuracy, but also set me straight about how chemistry *really* works. The accuracy of this book can be attributed to Bruce's help—any mistakes can be considered to be my fault.

The wonderful photos in this book were taken by Jannette Bloom, an up-and-coming young photographer. Thanks, Jannette!

I'd like to thank my wife Ingrid for making my life wonderful in so many ways. Why this amazing woman puts up with me, I have no idea. I'd also like to thank my parents, Matt, Cindy, Owen, my grandparents, my in-laws, and all of the other family members who have made my life immeasurably better.

Thanks to Marcello DiMare who made chemistry come alive for me. Thanks to Nancy Levinger for helping to convince me that I'd rather be a teacher than a chemist after all.

As if those weren't enough, thanks to the following friends who have supported me in a variety of ways: All of my students (past or present), my colleagues (past or present), Joy McManus, Jan Warner, Donna Delano, Pam Meier, Kevin Whelan, Cheryl French, Erika Popejoy, Maalox and Catnose, Brad Luther and Corinna Cincotta, Erik Luther, Rob Keil and Linda Williamson, Meikka Cutlip, Anita Scovanner Ramsey, Kat Wallace, the folks at the local post office (Carl, Bing, Debbie, Michelle, and Sonja), Gene the UPS guy, John the dog walker, and my neighbors for putting up with all the punk rock I blast on the stereo while writing.

Finally, a special thanks to you, the reader. Without you, I wouldn't have gotten such a big advance for writing this book. Ka-ching!

Trademarks

Part 1

Firing Up the Bunsen Burner

Welcome to chemistry! You're probably shaking in your boots, wondering what terrible things will be found in this book.

Don't worry! We're going to ease into this chemistry thing by learning about the guys who invented modern chemistry. After we've become familiar with these folks and the amazing things they had to say, we'll take a quick tour through some of the basic vocabulary of chemistry—things like the metric system, the scientific method, and the atom. By the time we're done with this section, I'm sure you'll feel a lot more comfortable with chemistry than you do now.

Okay … take a deep breath and turn the page.

Chapter 1

Why Study Chemistry?

In This Chapter

- What chemistry is good for
- Why you might hate chemistry
- Why you shouldn't hate chemistry
- The origins of chemistry
- How to learn chemistry

When most people think of the word "chemistry," they think of a sinister old man in a white lab coat giggling evilly over a bubbling beaker. This image probably comes from movies and television, which usually portray chemists as the creators of terrible monsters, world-destroying super-weapons, and insects that grow to abnormal sizes and terrify the residents of small towns in Texas. Fortunately, in recent years, the media has revised their former image of chemists—we're now sometimes depicted as senile rather than insane.

In any case, most people put off taking chemistry until the last possible minute, the same way most of us put off root canal surgery until flames start shooting out of our teeth. Don't worry if you're in the same boat. Historically, chemistry has been a source of stress since the times when it was first realized that mixing chemical A with chemical B formed a green powder that fended off witches. Fortunately, recent discoveries indicate that witches don't exist.

Why Do We Need Chemistry?

"Chemistry" is a hard word to define. Some chemistry textbook covers show pictures of bubbling flasks, suggesting that chemistry can be defined as "the study of how we can make things behave if we mess with it in the laboratory." Other chemistry books have pictures of huge molecules on the cover, suggesting that chemistry is defined as "the study of how we can cram atoms together to make big complicated structures." I've even seen a textbook cover that featured a multicolored squiggle. I have no idea what that says about the study of chemistry.

It seems to me that if we put these two definitions of chemistry together, we get a reasonable idea of what the subject actually entails. Chemistry can be defined as using our knowledge of how matter is put together and how it interacts with other matter to solve confusing problems.

The Mole Says

The best way to understand chemistry is to translate the big words into more easily understood terms. By translating scary 10-syllable words into easy 2-syllable words, you'll have a much easier time understanding the topic being studied.

Some of the confusing problems that you'll have to solve can be found in any chemistry textbook (including this one). A typical example:

> "What is the volume of 556 grams of steam at a temperature of 2300° Celsius and a pressure of 35.40 atmospheres?"

You're already painfully aware that these problems aren't important in our everyday lives. When will you ever need to find the volume of 556 grams of steam under the conditions indicated in the problem? Probably never.

So the obvious question is this: If I just told you that you'll never need to solve these problems in the real world, why do you need to learn chemistry?

The reason you have to learn how to solve problems like this is that they really do have applications in the real world. Knowing how to find the volume of 556 grams of steam under extreme conditions may not be something you'll be doing in the future, but you can bet that if you do anything scientific in the future, the law that allows you to answer this question will come in handy. Like basic arithmetic, basic chemistry is useful because it gives you the tools to solve real-world problems.

What Have I Gotten Myself Into?

At first glance, chemistry doesn't seem like it will be a lot of fun. As an example, flip through almost any introductory chemistry textbook, and you'll find confusing diagrams like Figure 1.1.

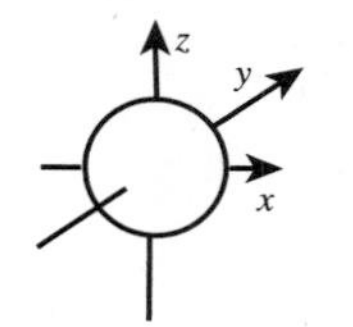

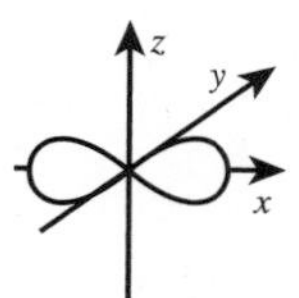

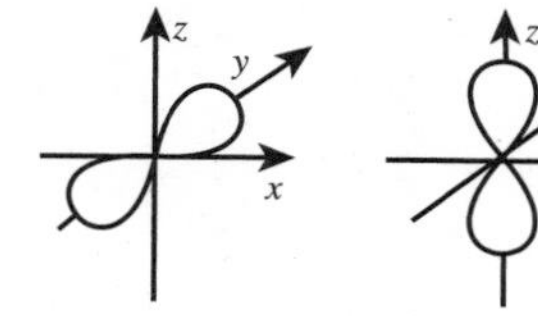

Figure 1.1

These unfriendly looking diagrams give us an idea of what atomic orbitals look like.

How the heck are you supposed to understand what this diagram means? Who came up with this stuff, anyway?

Relax. Take a deep breath. The reason you're nervous about taking chemistry is not because chemistry is difficult. The reason you're nervous is that you're trying to wrap your brain around every aspect of chemistry at once. Think back to the first time you learned to add. Wouldn't it have been terrifying to look in the back of the book to find a diagram explaining how to do long division? Take chemistry one step at a time and you'll do much better than if you confuse yourself with things we haven't discussed yet!

Bad Reactions

The best way to make chemistry confusing is to wait until the last minute to learn everything you need to know. By breaking the subject into small, easy-to-digest pieces, chemistry becomes much easier to understand.

This Stuff Is Really Fun!

Chemistry is fun! With a good grounding in chemistry, you'll not only understand how to balance an equation and discuss moles like a pro, but you'll also get a better understanding of how real-world phenomena work. For example, take the everyday fish tank filter (see Figure 1.2). Most of us don't give a lot of thought to how the filter works, choosing instead to scrub out the foam sponge and change the charcoal whenever the tank starts looking a little scummy.

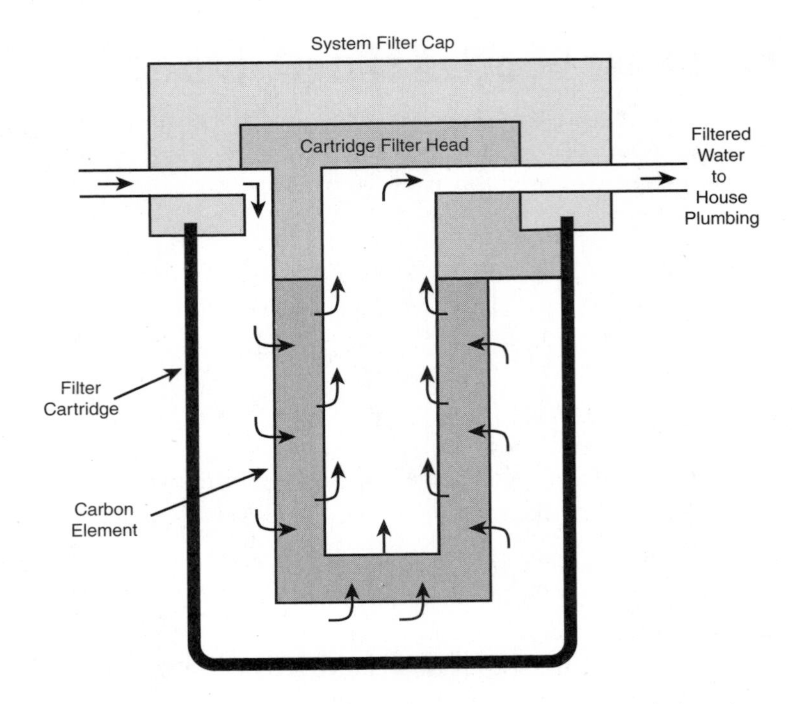

Figure 1.2

The foam rubber filters solid wastes, while the activated charcoal removes liquid wastes.

Though you wouldn't think so by looking at it, the filter uses sophisticated chemistry to keep your fish alive and happy. The first section of the filter contains a foam sponge that allows water to pass through undisturbed but picks up solid fish waste. From a chemist's standpoint, you've filtered the solids from a heterogeneous mixture. The second section of the filter contains activated charcoal, which serves to remove dissolved fish wastes. From a chemist's standpoint, the activated charcoal selectively *adsorbs* organic compounds, allowing inorganic materials to pass through undisturbed.

The Mole Says

When something is **adsorbed,** it has been stuck to the surface of a material. Absorption, with which you are more familiar, is when something has been soaked up into another material.

Okay, so this example won't wow the guests at your next get-together. However, you'll probably find that people are more interested in the science behind everyday objects than they're willing to admit, even if they do claim to hate chemistry.

Who Invented Chemistry?

If you haven't yet been convinced that chemistry is a worthwhile pursuit, you may be wondering who you can blame for its invention. Bad news—you'll have to travel back in time to punch the inventors in the nose, because chemistry has been around for thousands of years.

It's All Greek to Me!

Though it's not entirely clear when people started using chemistry, the first people to record their studies were the ancient Greeks. For this, as well as their many philosophical ponderings, students hate them to this day.

Ancient Greek scientists are primarily known for coming up with the idea of elements as well as early models of the atom. Unfortunately, the limitations of their technology kept them from getting an accurate idea of what these elements were and what atoms really looked like.

Turning Trash into Gold

For a very long time in the Middle Ages, chemistry was a mystical pseudo-science performed by alchemists whose goal was to turn cheap metals such as lead into gold using mysterious chemical processes. Typically, the works of the alchemists were mystical, involving spells and potions.

Though their science was a little flaky, the alchemists did keep the knowledge of the ancient Greeks alive while adding some touches of their own. Islamic alchemists in particular developed many of the laboratory techniques we use today, most notably the use of distillation to purify liquids.

> **Chemistrivia**
>
> One of the greatest alchemists was Jabir ibn Haiyan, who lived in the eighth century C.E. In addition to his quest to make gold, he wrote about dyeing fabrics, making fabric waterproof, and refining metals.

Chemistry Hits the Big Time

The first modern chemist was Robert Boyle (1627–1691). Though most famous for his work with gases, Boyle was also the first to disagree with the Greek idea of four elements in his book *The Skeptical Chymist*, published in 1661.

Despite his groundbreaking work, Boyle continued to believe that metals weren't really elements and that it would eventually be possible to convert one metal into another using chemical processes. Hey, even the greats sometimes strike out.

Modern Chemistry

Nowadays, chemistry has been converted from a quest to make gold into a big business with hundreds of thousands of chemists working worldwide. However, the quest to make valuable materials continues to be the driving force for modern chemistry.

One of the largest areas of chemical research today is the development of new pharmaceuticals. Because antibiotic resistance is a growing problem when treating many diseases, new drugs are continually being developed. The treatment of the HIV virus has been revolutionized by the use of protease-inhibiting medications. Organ transplants are made possible by the use of anti-rejection medications. Modern medicine simply wouldn't be possible without chemistry.

In fact, most of the stuff around your house benefits in one way or another from the practice of modern chemistry. The food you eat is colored, flavored, and preserved by various chemical additives. The cleaning supplies you use to keep your house from being closed by the health department are manufactured in industrial lots by large chemical firms. The bug killers you use to keep cockroaches from overrunning your kitchen are made in giant labs. Modern life simply wouldn't be possible without the use of chemistry.

Succeeding Where Others Have Failed

Even after reading this chapter, the big question in your mind may be, "Can I really learn chemistry?" The answer: With patience and time, anybody can learn chemistry. In my years of teaching chemistry I've never found anybody who, with a little bit of hard work and study, wasn't able to understand the basics. Of course, there are some things you can do to make this task easier on yourself:

- *Learn the vocabulary!* As mentioned earlier, chemistry is full of confusing, specialized terms. It doesn't matter how well you've memorized the material—if you don't understand what the words mean, you won't understand chemistry.
- *Learn processes, not facts!* A lot of people attempt to learn chemistry by memorizing the periodic table, the names and formulas of every chemical compound, and all the equations from the chemistry book. Some of these people learn chemistry, some of them don't, and some of them have heart attacks from all of that memorization. Chemistry becomes much easier if you learn how to solve problems, rather than memorizing the answers to every potential problem. In this book, we'll be discussing methods you can use to solve problems, rather than committing the atomic masses of the lanthanides to memory.
- *Slow and steady wins the race!* I certainly hope you bought this book sometime in the middle of a chemistry class, rather than the day before your final exam. Like most subjects, chemistry is much easier to understand if you take it in small, easily digested chunks. Remember, it took over 2,000 years to develop modern chemistry—you probably won't become a pro after a 32-hour study session.

- *Use common sense!* When you solve problems in chemistry, look at the answers to make sure they're right. For example, if you've found that you're 45 meters tall after doing a unit conversion, it's fairly certain you've made a mistake (unless you're the Jolly Green Giant).
- *Enjoy the scenery!* Though there are some really boring aspects to chemistry, there are also a lot of really neat things to learn. Think of chemistry as being like a long car trip—sometimes you have to endure the traffic in New Jersey before you can enjoy New York City.

Chemistrivia

When I took chemistry for the first time in high school, I barely passed after a great deal of effort. I know how you feel about chemistry and will help you get through the process as painlessly as possible (and with much better results than I had the first time through!).

The Least You Need to Know

- Chemistry is important because most of the other sciences use it as a tool for solving problems.
- A good knowledge of chemistry will allow you to understand how many common things work.
- Chemistry was developed over a very long period of time, but has only in the past two centuries been made into a real science.
- With patience and time, anybody can learn chemistry.

Chapter 2

Measuring Up

In This Chapter

- The metric system
- Performing unit conversions
- The importance of accuracy and precision
- Significant figures

We all know how to find the length of our thumb. All of us also know how to figure out how much we weigh (though some of us probably wish we didn't). These are simple everyday measurements involving very little specialized learning.

Of course, nothing is that easy when you take a chemistry class. For those of us used to thinking in "inches," "pounds," and "degrees Fahrenheit," we're told that we have to use the mysterious "centimeters," "kilograms," and "degrees Celsius." Not only that, but it's not enough to look at a ruler and just read off the length of our thumb—all of a sudden, we have to worry about the mysterious term "significant figures" and whether we've got the right number of them.

Calculations are worse—instead of putting numbers into your calculator and writing down the result, you now include only *some* of the digits, excluding others. Again, the concept of "significant figures" rears its ugly head in unusual and unpredictable ways. What's a chemistry student to do?

Not to worry. All of the above terms are simple to understand and use in the laboratory. The problem doesn't lie with the terms or with the ideas—the problem lies with the way that they're explained. In this chapter, you'll arm yourself with the necessary tools for taking and understanding chemical data.

The Metric System

You've probably been raised to believe that the grocery store is two miles away and that your height is 5'10'' (actual height may vary). The temperature outside is probably about 80° during the summer and 40° during the winter. Unfortunately, though each of these values is probably correct, they mean very little to a chemist.

Chemistry, like most of the sciences, depends on the International System of Units (called the SI system, from the original French "Systeme Internationale d'Unites"). The International System is just a fancy name for the "metric system," so you will probably see both used interchangeably.

Chemistrivia

In 1889, when the meter and kilogram were being defined, a problem arose: if you have two meter sticks and one is longer than the other, which is right? To solve this problem, a bar of metal and a chunk of metal were made of a platinum-iridium alloy and defined as being exactly one meter and one kilogram, respectively. The official meter bar and kilogram weight now reside at the International Bureau of Weights and Measures outside of Paris, France.

The SI system of units contains seven base units, shown in the following table.

SI Base Units

Quantity	Unit	Symbol
Length	meter	m
Time	second	s
Mass	kilogram	kg
Temperature	Kelvin	K
Electric current	ampere	A
Amount of substance	mole	mol
Luminous intensity	candela	cd

Sometimes, we find it's not very handy to use those units. For example, if we're trying to find the distance from your house to the northernmost point in Canada, the unit "meter" is difficult to use because of the lengthy distance. As a result, the metric system uses a series of prefixes that we can add to the previous units to make them easier to use. Below are the most commonly used prefixes.

Selected Prefixes for Metric Units

Prefix	Symbol	Meaning
giga	G	1,000,000,000 or 10^{9}
mega	M	1,000,000 or 10^{6}
kilo	k	1,000 or 10^{3}
deci	d	0.1 or 10^{-1}
centi	c	0.01 or 10^{-2}
milli	m	0.001 or 10^{-3}
micro	µ	0.000001 or 10^{-6}
nano	n	0.000000001 or 10^{-9}
pico	p	0.000000000001 or 10^{-12}
femto	f	0.000000000000001 or 10^{-15}

Let's say we live in Washington, D.C., and we're commuting to Baltimore, a distance of approximately 70,000 meters. Instead of saying that we had to travel 70,000 meters, we'd say that we had to travel 70 kilometers, or 70 km. Because "kilo" multiplies the unit by a thousand, 70 kilometers can be thought of as being $70 \times 1{,}000$ meters, or 70,000 meters. Similarly, if we've found that the length of our thumbnail is 0.02 meters, we can express this as 2 centimeters or 2 cm. Think of 2 centimeters as meaning 2×0.01 meters, or 0.02 meters.

You've Got Problems

Problem 1: Use metric prefixes to express the following measured values:

(a) 0.0000075 meters

(b) 25,000,000 grams

Converting Nonmetric Units to Metric Units Using the Factor-Label Method

Let's say that you've been stopped at the customs desk while entering a foreign country. After looking at your passport, the customs official seems to think you resemble the head of a major international drug cartel. To prove your identity, all you have to do is tell the customs official how tall you are in centimeters. If you can't prove your identity, the official will throw you in prison for 45 years of hard labor. What will you do?

At this point, you're probably thinking, "I hope the color of the prison overalls goes well with my shoes." However, it's possible to make this conversion with very little trouble. You just have to use the factor-label method, which is a simple way of converting between different sets of units. This method is described in the next few pages.

Single-Step Unit Conversions

To solve this problem, let's learn how to use the "factor-label" method to convert your height from inches to centimeters.

Let's say your height is 5'11", or 71 inches. The factor-label method allows us to solve this problem by following these steps:

Step 1

Write the unit you're trying to convert down on your paper:

71 in.

Step 2

Write a times sign after the unit you're trying to convert, followed by a straight, horizontal line. (Hey, this stuff isn't so bad after all!)

71 in. × ______________

Step 3

In the space below the line, write the unit of the number you're trying to convert.

In this case, the unit of the number you're trying to convert is "in." Write this below the line.

$$71 \text{ in.} \times \frac{\qquad\qquad}{\text{in.}}$$

Step 4

Write the unit of what you're trying to *find* on top of the line.

In our example, we're trying to find out how many centimeters tall we are. In this case, the unit is cm.

$$71\text{ in.} \times \frac{\text{cm}}{\text{in.}}$$

Step 5

Write the conversion factor in front of the units on the line.

Uh, oh. Problems. We don't know what a conversion factor is, which makes this an extremely difficult problem. Fortunately, the guy behind you in line taps you on the shoulder and explains that conversion factors are just numbers that allow you to convert from one unit to another. Even better, he knows that the conversion factor between inches and centimeters is "2.54 centimeters in 1 inch." As you turn to thank him, he melts back into the crowd.

def•i•ni•tion

A **conversion factor** is a number that allows you to convert one set of units to another.

The *conversion factor* tells us that there are " 2.54 centimeters in 1 inch," so let's write 2.54 in front of "cm." and "1" in front of "in.":

$$71\text{ in.} \times \frac{2.54\text{ cm.}}{1\text{ in.}}$$

Step 6

Solve the math problem you've written, making sure to cancel out any appropriate units.

In our example, the unit "in." cancels out, leaving us with "cm." as our unit:

$$71\ \cancel{\text{in.}} \times \frac{2.54\text{ cm.}}{1\ \cancel{\text{in.}}} = 180\text{ cm.}$$

You've Got Problems

Problem 2: Convert 160 pounds to kilograms. There are 2.21 pounds in one kilogram.

A hush falls over the crowd at customs. A sad look comes over the official's face as he signals his partner to put away the stun gun. You've made it!

Multi-Step Unit Conversions

At the end of your visit to this foreign country, a policeman stops you in the terminal, indicating that he has some routine questions he'd like answered. First, he asks how many fortnights you've spent in the country. When you answer that you've been in the country for 42 days, he frowns and insists on the answer in fortnights.

Fortnights? Who knows anything about fortnights? When you ask him the difference between days and fortnights, he grins and tells you that there are two weeks in a fortnight. What can you do?

Fortunately, the factor-label method can again come to your rescue. Let's go through the steps again.

Step 1

Write the unit you're trying to convert down on your paper.

42 days

Step 2

Write a multiplication sign after the unit you're trying to convert, followed by a straight, horizontal line.

42 days × ______________

Bad Reactions

A common mistake people make when doing unit conversions is turning multi-step unit conversion problems into one-step conversion problems. By making sure that one set of units cancels each other out during the problem, you minimize your chances of making this error.

Step 3

Below the line, write the unit of the number you're trying to convert.

$$42 \text{ days} \times \frac{\phantom{\text{days}}}{\text{days}}$$

So far, so good. No problems yet!

Step 4

Write the unit of what you're trying to *find* on top of the line.

Here lies the problem: If we write the unit that we're eventually trying to find, "fortnights," on the top of the line, we can't solve the problem because we don't know how many days there are in a fortnight.

However, we know how many weeks there are in a fortnight because the officer told us. Even better, we know that there are seven days in a week. To solve this problem, convert days to weeks. After we have finished this conversion, we can convert weeks to fortnights, using the conversion factor given to us by the policeman.

$$42 \text{ days} \times \frac{\quad \text{weeks}}{\quad \text{days}}$$

Step 5

Write the conversion factor next to the appropriate unit.

$$42 \text{ days} \times \frac{1 \text{ weeks}}{7 \text{ days}}$$

Step 6

Multiply these fractions together (with the calculator you keep with you while traveling to foreign countries), making sure to cancel out any appropriate units.

$$42 \cancel{\text{days}} \times \frac{1 \text{ weeks}}{7 \cancel{\text{days}}} = 6 \text{ weeks}$$

Now that we know how many weeks we have been in the country, we can convert weeks to fortnights using exactly the same method. This calculation should look like the one below.

$$6 \cancel{\text{weeks}} \times \frac{1 \text{ fortnight}}{2 \cancel{\text{weeks}}} = 3 \text{ fortnights}$$

That gives you a final answer of 3 fortnights. Once again, unit conversions have saved the day!

You've Got Problems

Problem 3: Convert 25 miles to meters. There are 1.6 kilometers in a mile.

The Mole Says

This calculation can be simplified by putting both steps together into the same calculation. The resulting calculation should look like this.

$$42 \cancel{\text{days}} \times \frac{1 \cancel{\text{weeks}}}{7 \cancel{\text{days}}} \times \frac{1 \text{ fortnight}}{2 \cancel{\text{weeks}}} = 3 \text{ fortnights}$$

Accuracy and Precision

For most people, the words *accuracy* and *precision* are interchangeable and mean that an act has been done well. However, in chemistry, there is an important difference between these seemingly similar terms.

Measured data are accurate when they are close to the true value. For example, if you weigh a pen on a balance and find that the mass is 8.8 grams, the balance is said to be accurate if the true mass of the pen is 8.8 grams. On the other hand, if the mass of the pen is really 4.5 grams, the balance is said to have very low accuracy.

Precision refers to how often a particular piece of data can be reproduced by the instrument measuring it. For example, if we weigh a pen three times and find that the mass each time is 8.8 grams, the balance is said to be precise. If the three measurements found that the pen weighed 6.7 grams, 7.9 grams, and 8.6 grams, the balance is considered imprecise.

> **def•i•ni•tion**
>
> **Accuracy** is when a measurement is close to the actual value of something, and **precision** is when a measurement gets the same value when repeated several times. Though accurate data are always precise, precise data may not always be accurate.

If you've ever played darts, you already have a good idea of the difference between accuracy and precision. When people play darts for the first time, they rarely hit the center of the dartboard, and the darts will fly all over the room, hit the wall, go through windows, and scare the dog. At this point, their darts are flying with neither precision nor accuracy.

After playing darts for a few hours, people frequently find that their darts are all grouped very closely, but that they are not in the center of the dartboard. We would say these darts were thrown with precision because they're in the same location, but also inaccurate because they're not in the center of the dartboard. Once people become good at throwing darts, the darts are frequently grouped tightly together at the center of the board. These darts are thrown with both precision and accuracy.

When collecting data in the lab, the instruments that we use have varying degrees of precision and accuracy. Inexpensive instruments such as bathroom scales give data that are rarely correct and frequently vary from measurement to measurement. Generally, we find that cheap equipment is neither precise nor accurate. However, pricey equipment is usually both precise and accurate because it repeatedly gives the right answer.

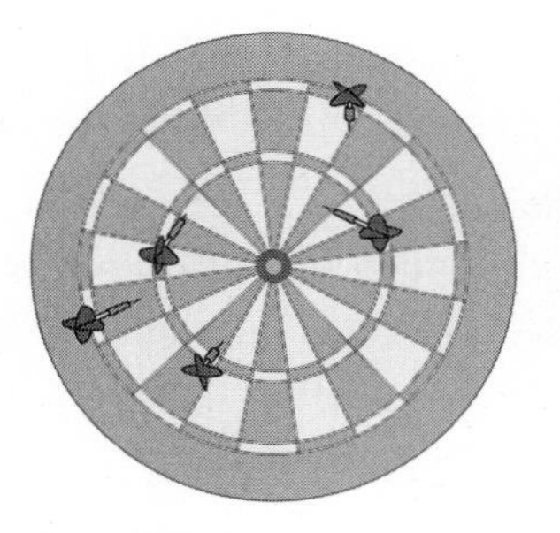 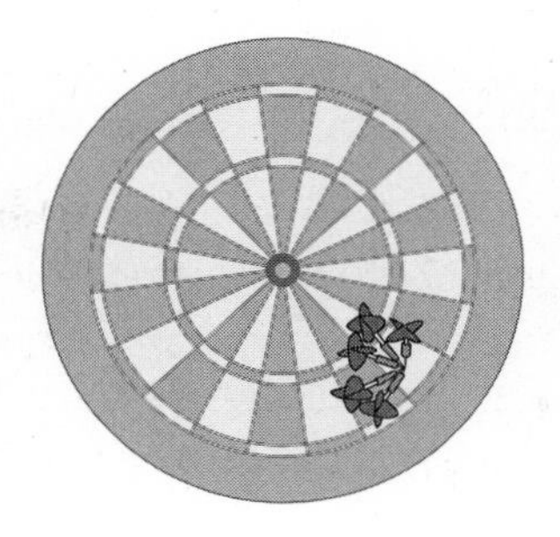 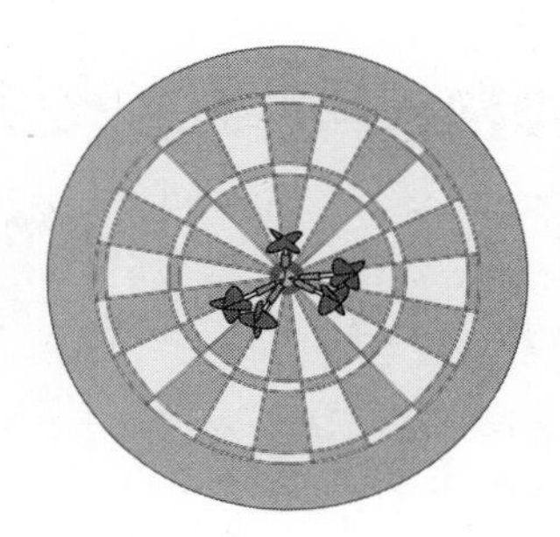

Figure 2.1

The first dartboard shows darts thrown with neither precision nor accuracy, the second shows darts thrown precisely but not with accuracy, and the third shows accurate and precise dart throwing.

Significant Figures

Now that we've learned a little bit about accuracy and precision, it's time to tackle the sticky subject of significant figures.

What Are They?

Let's imagine that I'm trying to find the density of a piece of foam rubber. Unfortunately, I don't have any equipment with me, so I'll have to just estimate its volume and mass. After careful deliberation, I determine that the *mass* of the foam rubber is somewhere around one gram, and the *volume* is about nine milliliters.

> **def•i•ni•tion**
>
> **Mass** is a measure of how much matter is present in an object and is usually measured in grams. **Volume** is a measure of how much space an object occupies and is measured in cubic centimeters (cm^3) or milliliters.

The density of an object is determined by dividing the mass of an object by its volume. Using the rough data I estimated above, I find that the density of the foam rubber is

$$\frac{1 \text{ gram}}{9 \text{ milliliters}} = 0.111111111 \text{ g/mL}.$$

Let's think about this for a minute. I told you I found both the volume and mass of the foam rubber by rough estimation. How, then, can I find the density to the nearest billionth of a gram per milliliter?

The answer is that I can't. If I have data that aren't very good to start with, I simply can't write my answer in such a way that it suggests the initial data were perfect.

Likewise, if I use a crummy piece of equipment to find the mass of an object, it would be odd to write the measurement to nine decimal places because the odds of the last nine being correct are slim.

def•i•ni•tion

Significant figures are the number of digits in a measured or calculated value that give us meaningful information about the thing being measured or calculated.

To avoid this problem, the idea of *significant figures* was devised. Significant figures are the number of digits in a value that give us meaningful information about the thing being measured or calculated.

Below is an example of how significant figures are used. Imagine that I've used a balance to find the weight of a paper clip. After measuring five times, I have collected the following data.

Measurement	Mass (grams)
1	1.1202
2	1.0911
3	1.0858
4	1.1418
5	1.1179

How heavy should I say the paper clip is? By examining the data above, we see that each measurement agrees that we have one gram—from this, we can conclude that this digit is probably valid. Looking at the tenths of a gram, it looks as if all of the measured numbers round to 1.1. After that, the digits appear random, meaning that they probably don't have any real meaning. As a result, we'd say that the paper clip weighs 1.1 grams. Since the answer has two digits, this measurement is said to have two significant figures.

It's important to write the correct number of significant figures when collecting data because it gives people a solid idea of how good your data are. To get the most out of your measurements, use the following rules when taking data.

1. When using analog instruments (instruments that aren't digital, such as a ruler), always write the number of digits that can be directly measured with the equipment, plus an extra digit that you estimate.

   Take a look at Figure 2.2. Normally, we don't spend much time thinking about the number of digits we use to write the length of a paper clip. However, to do it correctly, we need to write down the number of digits that can be directly measured, plus an extra digit we estimate.

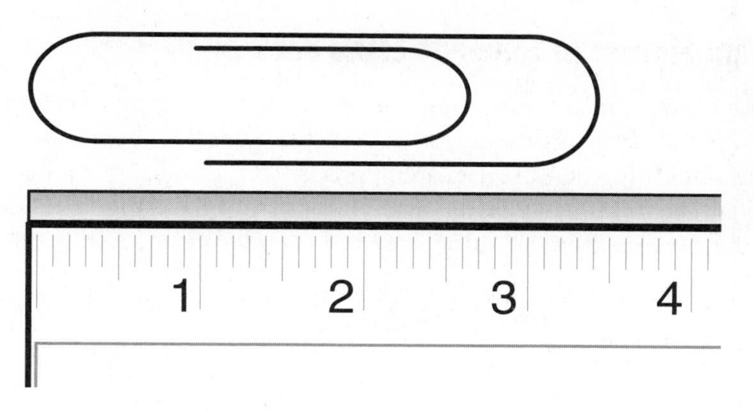

Figure 2.2

As you can see, this ruler has markings that show millimeters. As a result, we can estimate the length of the paper clip to the nearest tenth of a millimeter. In this case, the paper clip is 3.48 cm long.

Let's look at another example:

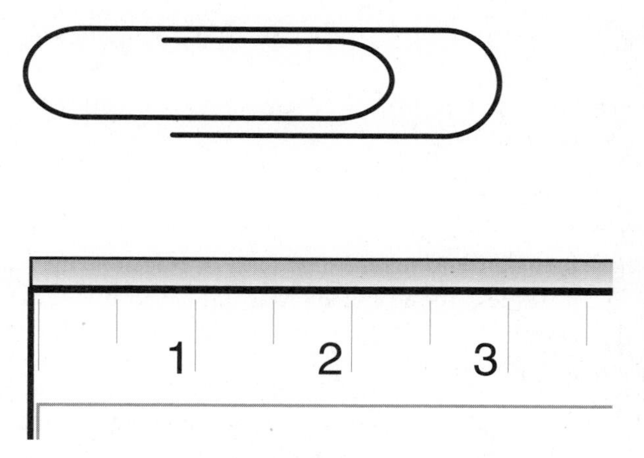

Figure 2.3

As you can see, the smallest markings on the ruler correspond to centimeters, which implies that we need to estimate the length of the paper clip to the nearest tenth of a centimeter. However, the paper clip appears to be exactly three centimeters long. As a result, we would say the paper clip is 3.0 centimeters long.

2. When using digital equipment (such as an electronic balance), just write down whatever is written in the screen.

Unlike analog equipment, it's impossible to make estimations using digital equipment. If an electronic balance tells us that the weight of a paper clip is 1.13 grams, we just have to trust that it's telling us the truth.

Determining Significant Figures in Someone Else's Data

You can tell a lot about how good a measurement is by the number of significant figures shown. For example, if poor-quality equipment has been used, there are usually only one or two digits, indicating that the data aren't precise. If high-quality equipment has been used, you may see four or five digits. The placement of these digits indicates not only how good the data are, but also if they've been rounded.

The following rules are used to determine the number of significant figures in a measurement.

Rule 1

All nonzero digits are significant. If the mass of a paper clip is recorded as 1.21 grams, all three of these digits give us useful information and are called significant figures.

Rule 2

Zeros between nonzero digits are significant. If the mass of a paper clip is 1.01 grams, all three of these digits are significant.

Rule 3

All zeros written to the left of all the nonzero digits are not significant. This is important for measurements with values less than one, such as 0.0023 grams. For this measurement, we would say that the two and the three are significant because of rule #1 above, but that the zeros are not significant.

Bad Reactions

Converting between two sets of units never changes the number of significant figures in a measurement. Remember, data are only as good as the original measurement, and no later manipulations can clean them up.

This seems counterintuitive. After all, don't those zeros give us useful information? Isn't useful information the point of significant figures? Yes, those digits do give us some useful information. To understand why they're not significant, imagine the following scenario.

I have found the weight of a paper clip to be 1.1 grams. This number has two significant figures. My boss has been bugging me lately about the quality of my data. As a result, I'll just convert the grams to kilograms, giving me a weight of 0.0011 kilograms.

The big question is this: did converting from grams to kilograms improve the quality of the data? It's pretty clear that it didn't—all we did was move a decimal place three spaces over to the left and add a prefix to the units. However, if we allowed those zeros to the left of the nonzero digits to be significant, we would go from a value with two significant figures (1.1 grams) to a value with five significant figures (0.0011 kilograms), which implies that the data improved. Since they didn't, we ignore the zeros on the left and treat them only as placeholders.

Rule 4

Zeros to the right of all nonzero digits are only significant if a decimal point is actually shown.

Let's consider three measurements that look identical. With one piece of equipment, I find that the weight of my computer mouse is 200 grams. With a second piece of equipment, I find that the weight of my computer mouse is 200. grams. With the third piece of equipment, I find that the weight is 200.0 grams. Are these answers the same?

Well, they look the same, but appearances can be deceiving. Here's what these numbers really mean.

Number	Meaning
200 grams	The "2" is the only significant figure in this value. This means that we've rounded our measurement to the nearest hundred grams. Pretty lousy data, huh?
200. grams	All three digits are significant because the decimal is shown. This means that the measurement has been rounded to the nearest gram. Not so bad.
200.0 grams	All four digits are significant because the decimal was shown. This measurement was rounded to the nearest 0.1 grams. Pretty good!

Rule 5

When using scientific notation, ignore the exponential part when finding significant figures. For example, if you're given the number 2.30×10^{-6} grams, ignore the "$\times 10^{-6}$" part and focus on the 2.30 part. This number has three significant figures: the 2, 3, and 0.

You've Got Problems

Problem 4: How many significant figures do the following measured numbers have?

a) 2.490 grams

b) 1010 grams

c) 0.01010 grams

Using Significant Figures in Calculations

Think back to the example where we found the density of foam rubber to be 0.111111111 grams/mL based on fishy data. Though we've discussed significant figures and what they mean, we still haven't discussed how to express calculations in significant figures.

As it turns out, the significant figure rules for calculations are fairly simple.

Rule 1

When multiplying or dividing numbers, determine which of the numbers has the smallest number of significant figures. The answer to the calculation should be written with the same number of significant figures as this number.

Let's go back to our density example. The mass of the foam rubber was about 1 gram, and its volume was about 9 milliliters. Because both of these numbers have one significant figure, the answer 0.111111111 g/mL is more properly written with one significant figure, or as 0.1 g/mL. By reducing the number of digits we write, we indicate to others that the data we used to find the answer were not that great to start with.

Rule 2

When adding or subtracting numbers, determine which of the numbers has the fewest decimal places. The answer to the calculation should be written to the same number of decimal places.

Let's say that we're trying to figure out how much of a chemical is in a beaker. If the weight of the empty beaker is 100.45 grams and the weight of the beaker containing the chemical is 105.761 grams, what is the weight of the chemical?

To solve this, we examine the two numbers to see which has the fewest decimal places. 100.45 rounds to the "hundredths" place and 105.761 rounds to the "thousandths" place, so our answer should be calculated to the hundredths place. As a result, our answer looks like this:

105.761 g – 100.45 g = 5.311 g, rounded to 5.31 g.

The Least You Need to Know

- The metric system is the fundamental set of units for chemistry.
- The factor-label method enables us to do unit conversions.
- Accuracy is a measure of how close experimental data are to the actual value, and precision is a measure of how often you can get the same answer.
- Significant figures are important because they let other people know how good our data are.

Chapter 3

The Scientific Method

In This Chapter

- What the scientific method is, and what it isn't
- Learn how to use the scientific method to solve problems
- Analyze experimental errors

My car doesn't work quite right. About six months ago, I found that whenever I made a high-speed turn to either the left or the right, the steering wheel began to vibrate uncontrollably and a loud grinding noise came from underneath the car. When I told my wife that I needed to do something to fix the problem, she gave me a doubtful stare and said that if the car vibrated when I made high-speed turns, maybe I should drive more slowly. Then she mentioned the six speeding tickets I had received in the past 10 years, and indicated that perhaps slower driving could kill two birds with one stone.

I took the car to the shop. After a quick look under the car, the mechanic said that the car looked okay to him and that maybe I should consider driving more slowly. At the next shop, the mechanic spent an hour test driving and examining the bottom of the car, looking for the problem. He came up with some suggestions for what might be the problem and offered to make my car "as good as new" for a mere $600, though it would probably be cheaper to just drive more slowly in the future. Then he asked whether I would be paying for the inspection by check or credit card.

How does all of this relate to chemistry? When solving problems in chemistry, it's important to have a systematic method for finding solutions. Otherwise, we're just left trying one thing after another, without reaching a satisfying conclusion.

Incidentally, I now drive more slowly.

Why Use Any Method at All?

Let's go back to the moment that my car started to act funny. The sounds and vibrations told me that there was probably something wrong with the steering, so my first step might have been to replace the steering wheel. If that didn't work, maybe I'd replace the front tires. If that didn't work, maybe I'd get a new front axle. This cycle might have kept going on and on until either the problem was fixed or I ran out of money.

Obviously, changing things at random isn't an ideal way to solve this problem. Random guesswork is an ineffective way to get things done because there's no systematic way to figure out whether you're on the right track or if what you've done has even had any effect on the problem. In my example, I would probably have gotten the car fixed after making some change or another, but it would be a lot more trouble than if I'd let somebody who knows about cars solve the problem.

Using the Scientific Method

The method that scientists use when solving problems is called, creatively enough, the *scientific method.* The scientific method gives scientists a framework with which to pursue their investigations.

The Steps of the Scientific Method

When presented with a problem, scientists (ideally) use the following steps of the scientific method to find a solution.

def•i•ni•tion

The **scientific method** is a systematic method of solving problems based on experiments and observations.

Step 1

Purpose: This is the thing we hope to achieve by doing an experiment. For example, when my car started vibrating, my purpose was to make the vibration stop. For scientists, the purpose might be "to come up with a new, more effective antibiotic" or "to improve the odor of cat litter."

Step 2

Hypothesis: This is an "if, then" statement that describes how we intend to solve the problem. The hypothesis in my example was, "If I take my car to the auto mechanic, then he'll make the vibration stop."

Hypotheses are often based on past observations. It's been my experience that when I take my car to the shop, the mechanics usually do a pretty good job of figuring out what the problem is and restoring the car to health. Subsequently, it was not unreasonable to think that the same thing would happen when I took the car in for this problem.

Step 3

Devise an experiment to test the hypothesis. Clearly, a hypothesis isn't helpful if there is no way to verify it.

Experimental design is a difficult process that may take years to perfect. One of the biggest problems scientists have with writing experiments is keeping them within realistic cost parameters. Scientists have bills to pay, and if the experiments are too expensive, they can't be done. Fixing my car by replacing random components falls under this category. Other problems with devising a procedure for an experiment include limitations in laboratory equipment, lack of expertise in the process being studied, time limitations, and safety requirements. Even though scientists have fancy lab coats and big research budgets, there are things that even they can't do!

Step 4

Data Collection: While performing an experiment, it's important that the data be collected in such a fashion that the experimenter is able to make the most of it.

What are data? This is a big question that most of us never really think to ask because we usually assume that we know what data are and how to get as much of them as possible from an experiment. In this, we're usually wrong.

There are two types of data—*quantitative* and *qualitative.*

- Quantitative data are what we're usually used to taking. The word "quantitative" is based on the word "quantity," and it stands for any data that involve a numerical measurement. Whenever you take a temperature, measure the amount of time it takes for something to happen, weigh something, and so on, it qualifies as quantitative data. For most experiments, it's obvious what quantitative data should be taken.

With my car example, quantitative data may include the amount of time it took for the mechanics to look over my car (30 seconds for the first mechanic, 1 hour for the second mechanic), how much money it cost to get the car looked at ($15 for the first mechanic, $65 for the second mechanic), or the number of times I was told "maybe you should just drive more slowly" (one time for the first mechanic, six for the second).

- Qualitative data refer to any observations that don't involve numbers. This is the sort of thing we usually observe every day, such as "that guy sure has a big nose" or "my cat smells funny today." Whenever we do an experiment, it's important that we write down everything that happens because we can never really be sure what's important until it's time to analyze the data. Sometimes, things that seem worthless hold the key to an experiment.

def•i•ni•tion

Quantitative data are any measurements that involve numbers, such as weights, lengths, times, or temperatures. **Qualitative data** consist of observations that can't be expressed as numbers, such as "the beaker started to vibrate," or "the firefighters were mad that they had to come out to my lab again."

For the example involving my car, the qualitative data would include things like, "The first mechanic gave me a funny look when I told him I didn't feel like driving more slowly" or "The second mechanic told me that he wasn't sure it was a good idea to fix my car knowing that it would only help me to drive more recklessly in the future."

Step 5

Data Analysis: This is where you examine the data to see if you were successful in solving the problem. The actual method of data analysis will vary from experiment to experiment, but what all data analysis sections have in common is that when you're done, you'll have a good idea as to whether or not the experiment you performed has solved the problem.

In my case, I found that although I took the car to several mechanics, the steering continues to make a funny noise when I make high-speed turns. As a result, my experiment was a failure, and I have to try something else if I'm going to keep making high-speed turns.

Step 6

Wash, Rinse, Repeat: If the experiment worked, you're done and can make the conclusion that whatever you did solved the problem. If not, then it's time to go back and try the whole thing again with a revised hypothesis. For my example, I've tried testing the new hypothesis, "If I quit taking turns so quickly, then the vibration will stop" with good results. Of course, I'd like to be able to drive my car more quickly, so I may revisit my original problem at some point in the future.

Chemistrivia

Real scientists sometimes spend years trying new approaches to solving the same problem. In science, success takes a long time to achieve—many problems are never solved at all. For example, it was reported in the mid-1980s that it was possible to perform nuclear reactions under room temperature conditions (called "cold fusion"). The reports of these reactions were later proved to be false, though some people continue to work on cold fusion to this day.

What About Dumb Luck?

Throughout the history of science, there have always been examples of people who made huge advances through dumb luck rather than methodical problem solving. Some chemistry textbooks like to pretend that the scientific method is the only way to make new discoveries, but the fact of the matter is that many new discoveries are made accidentally. As the saying goes, "chance favors the prepared mind." Here are two examples.

- The case of the fogged film. In 1895, a German scientist named Wilhelm Conrad Roentgen was doing research with cathode ray tubes to determine how far the rays could travel outside of the tubes. Weirdly, he found that the rays caused a screen on the other side of the lab to start glowing. This was particularly surprising because cathode rays were known to have a limited range. After some experimentation, he found that these new rays (called x-rays because they were so mysterious) could be used to take pictures of the bones in his wife's hand.
- The case of the dead bacteria. In 1928, Alexander Fleming was hard at work growing bacteria in Petri dishes. One afternoon, he decided to examine some of his bacteria cultures before washing them down the drain. To his surprise, he found that a mold had started growing on one of them, causing the bacteria to die. After some experimentation, he found that this mold (part of the penicillin family) was able to kill many different varieties of bacteria, starting a revolution in medicine.

In both cases, the scientific method played no part in the initial discovery of two very important scientific phenomena. However, you've probably noticed that both of these examples include the phrase "after some experimentation." In these cases, though the scientific method wasn't used to make the initial discovery, it was important in allowing the scientists to understand their discoveries. This is the greatest strength of the scientific method.

Using the Scientific Method to Solve Problems

Whether you know it or not, you probably use the scientific method every time you're trying to solve a problem. You may not write out all the steps, but the steps are familiar to you if you look for them.

A perfect example from my life: My shoelace broke.

- Purpose: To get a new shoelace.
- Hypothesis: If I look in my sock drawer, then I'll find a replacement shoelace.
- Experiment: Open the drawer and see if it contains a shoelace.
- Data Collection: No shoelace.
- Data Analysis: When I looked in the drawer, I didn't find a replacement shoelace.
- Repeat the experiment. If I look in my (desk drawer, wife's sock drawer, toolbox, file cabinet, etc.), then I'll find a replacement shoelace.

When I was looking for my shoelace, I didn't think, "My hypothesis is that …." I just figured that my sock drawer would be a good place to look. As this example illustrates, the scientific method is something we all use every day, whether we're conscious of it or not.

Error Analysis

One of the hardest things to do in science is to admit that we're wrong. Unfortunately, all of us make mistakes every time we perform an experiment. This isn't to say that we make huge mistakes while performing experiments. In fact, most of these mistakes are so small that we don't even notice them. After all, nobody's perfect.

Types of Error

Two kinds of error are recognized in the collection of data—systematic error and random error.

Systematic error (also known as determinate error) occurs when there's something wrong with an experiment that causes the values to be skewed by the same amount every time. For example, whenever my wife measures her weight on the bathroom scale, her weight comes out three pounds higher than her actual weight (or so she says). Though she's forbidden me to give numerical examples for the purpose of this book, her weight consistently reads three pounds higher whether she's been on a diet or been eating Oreos by the handful. As a result, she knows to compensate for this by always subtracting three pounds from whatever the bathroom scale says. Systematic error can be caused by almost anything: equipment failure, human error, the monsters living in the lab, etc.

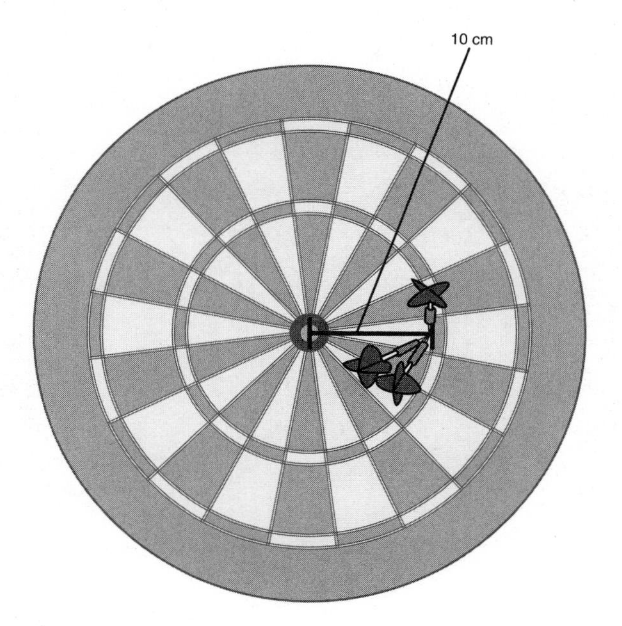

Figure 3.1

To compensate for the systematic error in my dart-throwing, I'll need to throw 10 centimeters to the left to hit the bullseye.

Bad Reactions

When our results don't prove our hypothesis, it's tempting to write it off as experimental error and say that the hypothesis has been proved anyway. One of the most difficult things for scientists to do is to retest their results several times with different methods before sharing them with anybody. The goal of this retesting is *not* to *prove* their data—the goal is to *disprove* it. If their data still hold up to the test, only then can the data be considered reliable.

def•i•ni•tion

Systematic error is when you get the same mistake every time you perform a measurement, and **random error** is when the mistake varies randomly. It's much easier to compensate for systematic error than for random error.

The other main type of error is *random error* (a.k.a. "indeterminate error"). Random error is exactly what it sounds like: error that can't be predicted or compensated for. For example, if I jump on the bathroom scale and find that my weight is 210 pounds the first time, 212 pounds the second time, and 213 pounds the third time, I can safely assume that the differences are caused by random fluctuations in how the scale works and not by a systematic error or actual fluctuations in my weight. As a result, my weight is probably close to the average of the three, or about 212 pounds.

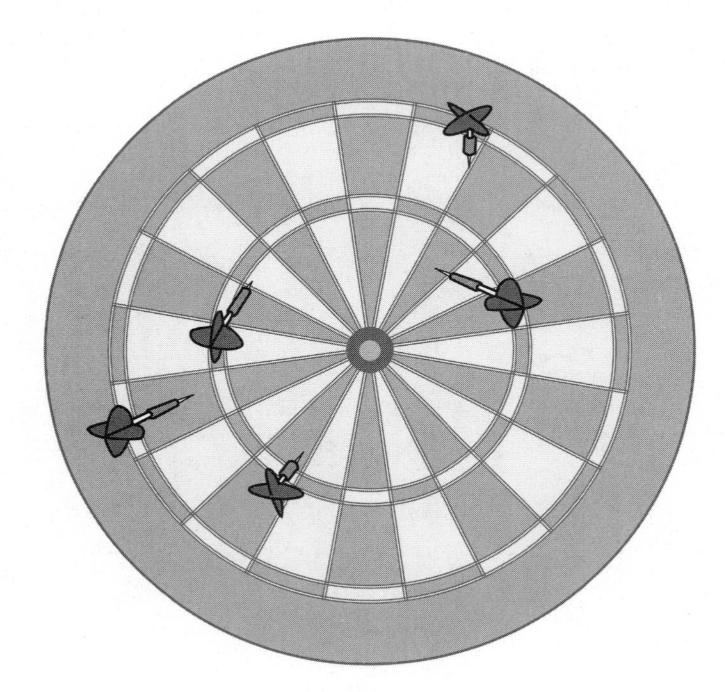

Figure 3.2

This dartboard shows random error that can't be compensated. In experimental settings, it's usually assumed that averaging a series of measurements approximates the actual value.

Random errors can be caused by anything from lousy equipment to a sloppy lab procedure. In the case of a friend of mine who works with microscopes sensitive enough to measure individual atoms, the vibrations from people walking outside the building are enough to make his data meaningless. As a result, he does most of his work at night.

Quantifying Error

As you might imagine, it's very difficult to measure exactly how much error is present in an experiment. After all, if you've got a systematic error, you may not know it's there at all. Random errors are also difficult to quantify because they're, well, random.

When working with experimental data, we usually assume that the last significant figure has a little bit of uncertainty in it. For example, if I find that the mass of a paper clip is 1.12 grams, I might write that as "1.12 +/– 0.01 grams," to signify that the last digit may be either a "1" or a "3" because of experimental error. In Chapter 11, we'll learn to quantify experimental error based on how much of a chemical compound is produced during a reaction.

The Least You Need to Know

- The scientific method is a methodical way of solving problems in science and in everyday life.
- The scientific method enables us to take full advantage of accidental discoveries.
- Experimental error can throw off the results of an experiment, but can sometimes be compensated for.

Chapter 4

The History of the Atom

In This Chapter

- Atoms, as perceived by the Greeks
- Dalton's laws
- The "plum pudding" model of the atom
- Rutherford's gold foil experiment
- Isotopes

Most of us think we have a pretty good idea of what atoms are. We know that they're tiny and that they're very important in chemistry. However, let's play a game. Define "atom" for me. You have 10 seconds.

Time's up! Let's hear it. What do you mean it's harder to define than you thought?

Believe it or not, the simplest ideas in chemistry are some of the hardest to articulate. If defining "atom" is giving you trouble, don't feel stupid. After all, it took science more than 2,000 years to get a good idea about what the atom looks like. Even now, the atom is still mysterious to us.

As a result, it's pretty hard to understand the atom all at once. Instead of diving headfirst into the deep end of quantum mechanics, we're going to float inside a big goofy inner tube in the shallow end.

What's an Atom?

An *atom* is the smallest chunk of an element with the same properties as larger chunks of that element. Let's imagine that we have a big block of gold. If we use a saw, we can cut the block in half without too much trouble. If we cut it again, we get an even smaller piece. Let's imagine that we repeat this process for a very long time. Eventually, we end up with a piece of gold so small that there's no way to break it into smaller parts. This tiny chunk is an atom.

def•i•ni•tion

An **atom** is the smallest chunk of an element that still has the properties of that element. For this reason, atoms are usually referred to as "the building blocks of matter."

Some of you might be asking yourselves, "Aren't protons, neutrons, and electrons smaller than atoms?" Of course they are! Remember: One of the things that defines an atom is that it has the same properties as a larger quantity of the same element. You can break an atom into even smaller pieces (it's tough!), but these pieces have different properties than the element you started breaking up in the first place.

Chemistrivia

We've mentioned that atoms are small, but exactly how small are they? To give you an idea, there are 6.02×10^{23} (that's 602 billion trillion) hydrogen atoms in one gram. As a result of this size, we usually don't count atoms separately but in larger groups called "moles" (see Chapter 11).

Older Atomic Theories

Let's take a look at various atomic theories, starting with the Greeks.

The Greeks Invent the Atom

The first people we know of who started thinking about the nature of matter were the ancient Greeks. The atomic theory was first created by Democritus and Leucippus, who concluded that when matter is divided, eventually the result is indivisible particles called atoms.

The nature of these atoms was not entirely clear to the Greeks because they had no way of testing their theories. As mentioned in Chapter 1, Plato believed that there were four elements (fire, air, earth, and water) that were the building blocks of all matter. He also believed that each of these elements had a unique shape, based on shapes called the Platonic solids.

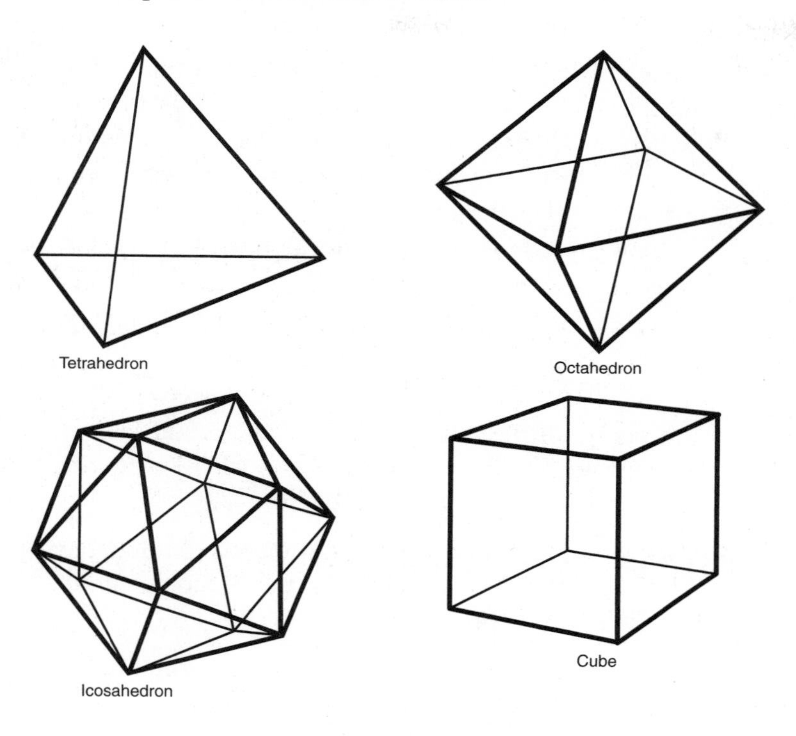

Figure 4.1

According to the ancient Greeks, atoms of fire were tetrahedral (four-sided), atoms of air were octahedral (eight-sided), atoms of water were icosahedral (twenty-sided), and atoms of earth were cubic (six-sided).

Despite the complete lack of evidence, this theory reigned for a very long time. It was a common belief during the dark ages that the Greeks knew everything there was to know, and this belief kept much of their knowledge alive.

Some Random Observations

For nearly two millennia, nobody really did any experiments to discover the nature of matter. In the late eighteenth century, this all changed with the discoveries of Antoine Lavoisier. Lavoisier found that the weight of the products of a chemical reaction was the same as the weight of the reactants. This is now known as the *law of conservation of mass.*

Chemistrivia

Unfortunately for Lavoisier, the French revolutionary government was not as impressed with his work as we are today. To fund his research, he worked as a tax collector, resulting in an unpleasant trip to the guillotine in 1794. Rejecting his appeal, the court commented that "The Republic has no use for savants."

To us, the law of conservation of mass is old news. After all, it seems intuitive that if we make a sandwich with 100 grams of ham and 50 grams of bread, the final weight of the sandwich will be 150 grams. However, let's look at an example that confused people back in the old days.

Imagine that your significant other has dumped you and you're burning all of his or her pictures. When you start the bonfire, the total weight of the pictures is 250 grams. However, when you weigh the ashes afterward, you find that they weigh only 50 grams. It looks as if matter was destroyed, perhaps turned into the energy that created the fire.

def•i•ni•tion

The **law of conservation of mass** states that the weights of reactants in a chemical reaction are the same as the weights of the products. No matter what chemical changes may occur, matter is neither created nor destroyed.

We know now that this isn't the case. When you burn something, the ashes account for some of the weight, while soot, smoke, and the other vapors given off account for the rest of it. The weight is still around, but it was moved to a new location where it can't be seen (in this case, the atmosphere).

Around the same time as Lavoisier's experiments, Joseph Proust discovered that when you analyze a chemical compound, the ratio of the elements in the compound always stays the same. This rule is called the law of definite composition.

This discovery created quite a stir. Previously it was thought that if you combined two elements, they would form a chemical compound in a ratio that depended on how much stuff you put together. For example, if you added two parts hydrogen to one part oxygen, you would make water with a formula of H_2O and if you added three parts hydrogen to one part oxygen, the resulting water would have a formula of H_3O. The law of definite composition made people realize that whenever you see a chemical compound, it always has the same formula.

The final controversial law was the law of multiple proportions. It states that when two elements combine to form more than one chemical compound, the ratio of the masses of one element that combines with a fixed mass of the other element can be expressed as a ratio of small, whole numbers.

What the heck does this mean? Let's use the example of water (H_2O) and hydrogen peroxide (H_2O_2). In both of these compounds, there are two atoms of hydrogen that weigh a total of two amu (don't worry, we'll explain the "two" part in Chapter 11 and the "amu" part in a few pages). In water, there is one atom of oxygen—this single atom of oxygen weighs 16 amu. In hydrogen peroxide, there are two atoms of oxygen, weighing a total of 32 amu."

The law of multiple proportions says that because the amount of hydrogen is the same in both compounds, when we compare the mass of oxygen in water to the mass of oxygen in hydrogen peroxide, the ratio should be a whole number. We see from the following that this ratio is 2.

Amount of oxygen in H_2O_2 = 32 amu

Amount of oxygen in H_2O = 16 amu

The ratio of oxygen in these two compounds is: 32 amu / 16 amu = 2

John Dalton Puts It All Together

In 1808, John Dalton used the existing knowledge about atoms to devise a series of rules describing the behavior of atoms, referred to as Dalton's Laws.

- *All matter is made of tiny, indestructible particles called atoms.*
- *All atoms of a given element are identical.* For example, if you purify gold by two different methods, each atom of gold from either method will have the same chemical and physical properties.
- *Atoms of different elements have different properties.* Some elements may have similar properties as one another, but no two elements have exactly the same chemical and physical properties.
- *Atoms are neither created nor destroyed in chemical reactions; they obey the law of conservation of mass.* When chemical reactions take place, only the arrangement of atoms changes, not the weight.
- *Atoms of different elements form compounds in whole number ratios.* For this reason, chemical compounds have formulas such as H_2O, not $H_{21}O_{13}$.

Nowadays, we know that some of these rules are true and some are false. Atoms aren't indestructible because we can break them apart in nuclear reactions. Atoms of the same element don't always have the same properties. Overall, Dalton didn't do so badly, considering that he hadn't actually seen any atoms directly.

J. J. Thomson: A Man and His Dessert

In 1897, British physicist J. J. Thomson performed an experiment in which voltage was applied across two wires (called electrodes) in a vacuum tube. As a result of this applied voltage, a glowing beam of particles was observed to travel from the cathode (the negative electrode) to the anode (the positive electrode). Because these light rays originated at the cathode, they were called, creatively enough, cathode rays.

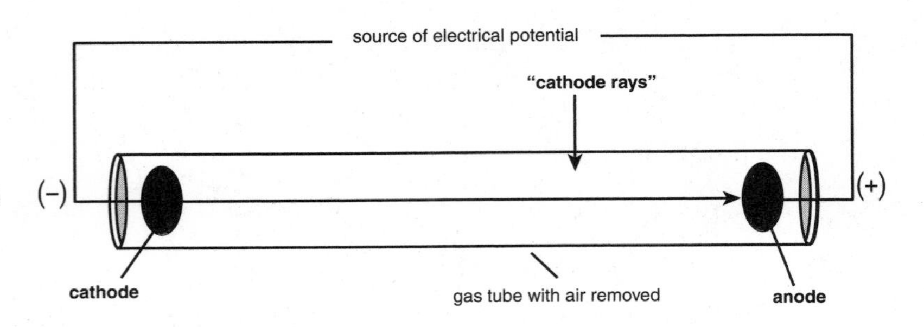

Figure 4.2

In the cathode ray experiment, Thomson was able to verify the existence of the electron characterize its mass, and determine that it had negative charge.

The Mole Says

Charged particles tend to move away from particles with the same charge and toward particles with the opposite charge. When the cathode rays bent away from the negative pole of the magnet and toward the positive pole, this rule caused Thomson to realize the cathode rays were negatively charged.

Through the use of magnets, Thomson was able to deduce that these rays contained particles with a negative charge. In his experiments, he placed the positive pole of one magnet on one side of the cathode ray tube and the negative pole of another magnet on the other. When the cathode ray tube was turned on, the ray was deflected away from the negative pole and toward the positive pole. This indicated to Thomson that the cathode rays consisted of particles with negative charge.

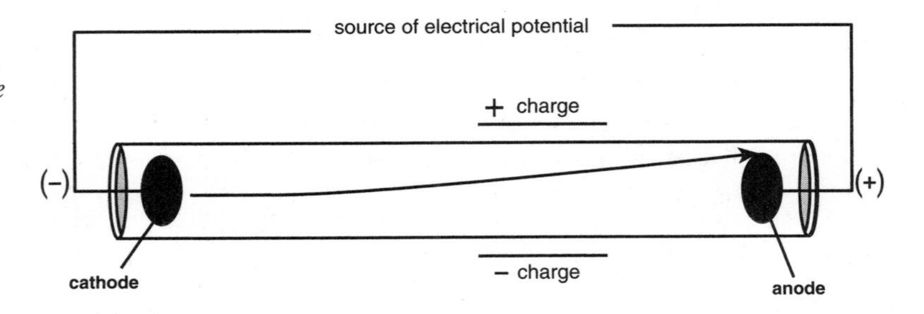

Figure 4.3

Because the cathode rays were deflected toward the positive pole of the magnet and away from the negative pole, Thomson determined that electrons have negative charge.

During the course of his experiments, Thomson determined that the mass of electrons was very tiny when compared to the mass of the overall atom. This led him to believe that electrons are very small and light when compared to whatever contained the positive charge in an atom. As a result, he devised the "plum pudding" model of the atom. In the "plum pudding" model, Thomson asserted that the atom was one big cloud of positive charge (the pudding), with very small particles of negative charge (electrons, the plums) embedded in it. Because not many people are familiar with plum pudding, a chocolate chip cookie is probably a more modern example.

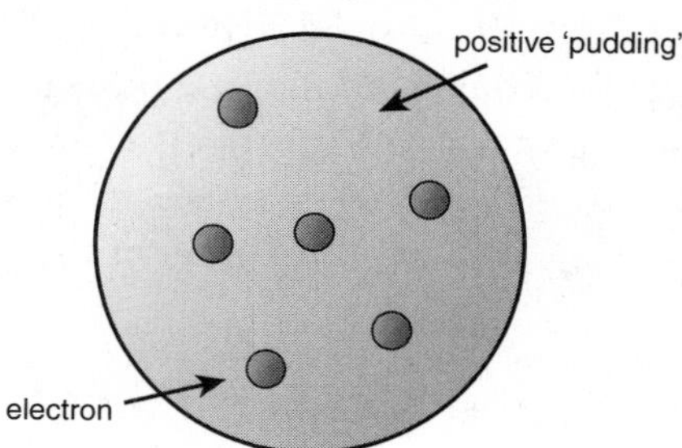

Figure 4.4

Named after his dessert, the plum pudding model portrays the atom as a big ball of positive charge containing small particles with negative charge. In this picture of a cookie, the dough represents the positive charge and the chips represent the negatively charged electrons.

Today, historians believe that had Thomson been diabetic, the course of atomic theory would have been vastly different.

Rutherford and His Alpha Particles

When radioactivity was discovered in 1896, lots of scientists decided that it was the great new thing to study. During the course of the next decade, scientists studied the decay of radioactive elements into smaller, energetic particles. In time, three types of radiation were discovered: Alpha radiation, which consists of helium nuclei with a +2 charge; beta radiation, which consists of electrons; and gamma radiation, which is very energetic light. (Don't worry—we'll discuss radiation in Chapter 26.)

One of the big researchers in the field of radiation at the time was Ernest "Radioactive Man" Rutherford. One day in 1907, Rutherford was messing around in the lab, shooting alpha particles at a very thin piece of gold foil to see how they would be deflected by the cloud of positive charge postulated by Thomson. Though most of the particles shot straight through as expected, some of the particles were deflected at very large angles. Some of the alpha particles even flew back toward the radiation source. This was definitely not what Rutherford expected.

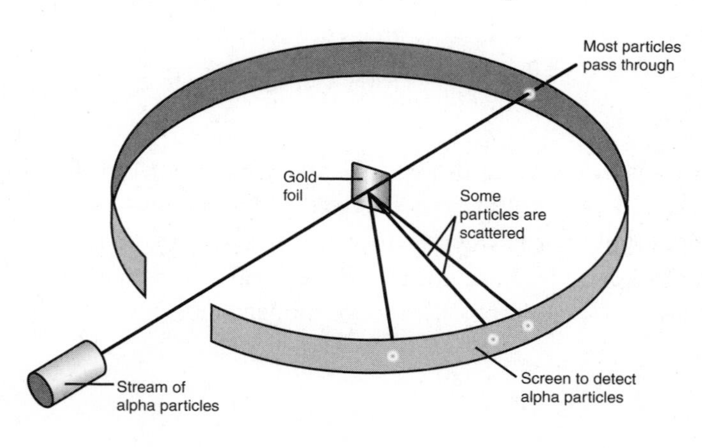

Figure 4.5

When Rutherford shot alpha particles at a thin piece of gold foil, he found that while most of them traveled straight through, some of them were deflected by huge angles.

Rutherford was eventually able to explain his discovery by coming up with a non-dessert related model of the atom. Instead of having all of the positive charges exist as a big heavy cloud, Rutherford theorized that the positive charge was concentrated in the "nucleus" of the atom, and that the negatively charged electrons floated around the nucleus. According to this model, most of the atom was empty space. The accompanying figure shows Rutherford's nuclear model of the atom.

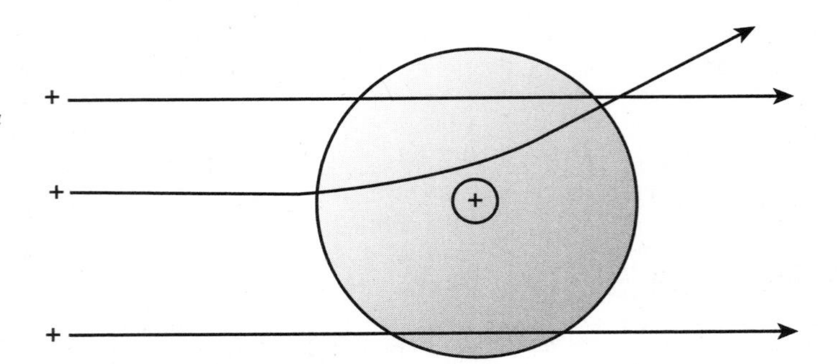

Figure 4.6

Rutherford believed that when positively charged alpha particles passed near the positively charged nucleus, the resulting strong repulsion caused them to be deflected at extreme angles.

Get to the Point, Already: What We Believe Now

We now know that the fundamental parts of an atom are the protons and neutrons, located in the nucleus, and the electrons, located in "orbitals" outside the nucleus. Take a look at this handy table.

Particle	Location	Mass	Charge
proton	nucleus	~1 amu	+1
neutron	nucleus	~1 amu	0
electron	orbitals	1/1836 amu (~0)	-1

To fully understand this table, let's discuss some atomic vocabulary.

- Orbitals are where the electrons reside in an atom. We're going to discuss this in great length in Chapter 5.
- The nucleus of the atom is where all of the positive charge and most of the mass of an atom is located. Think of it as the atom's creamy filling. The neutrons are also found in the nucleus, but they have no charge.

- The term "amu" stands for atomic mass unit. Because nuclear particles are tiny, it doesn't make much sense to discuss their weight in kilograms (protons and neutrons both weigh approximately 1.67×10^{-27} kg; electrons weigh approximately 9.11×10^{-31} kg).

The number of protons in an atom determines what element is present. For example, if an atom has one proton, it's hydrogen, regardless of how many neutrons or electrons it has. Atoms with the same number of protons and electrons are always neutral because the positive and negative charges cancel each other out.

Some of you may be wondering why atoms need neutrons at all. Neutrons are like my lazy Uncle Bob. Uncle Bob isn't really much of a nuisance—he just sits around his house drinking beer and watching TV. What good is Uncle Bob, anyway?

As it turns out, my aunt and her children would probably kill each other if it weren't for Uncle Bob. Though Uncle Bob usually just sits around watching *The Jerry Springer Show* and *Days of Our Lives*, his purpose becomes clear whenever my aunt and cousins start fighting with each other. When this occurs, Uncle Bob springs into action with his trademark phrase:

> "Shut up, all of you! I'm trying to watch *Oprah*!"

With that, peace is restored to the household.

In the same way, neutrons keep the nucleus from falling apart. Consider that the protons in an atom all reside next to each other in the nucleus. If you think about it, this isn't the most stable arrangement in the world because the positive charges repel each other. Fortunately, the neutrons are present to separate the protons and keep them from repelling the nucleus into oblivion. As far as I know, they don't watch *Jerry Springer*, but as I mentioned earlier, the atom still contains many mysteries.

Isotopes, Isotopes, Rah Rah Rah!

If you've spent any time at all studying atoms, you've probably bumped into the term "isotope." Let's figure out what that means.

Why Isotopes Exist in the First Place

Let's go way back to Dalton's laws. One of Dalton's beliefs was that all atoms of the same element have the same chemical and physical properties. As mentioned earlier, this isn't entirely true.

As it turns out, one of the ways that atoms of the same element may differ is in how much they weigh. Atoms of the same element that have different weights are called *isotopes*.

But why can there be more than one possible number of neutrons in an atom? Let's go back to the example of my Uncle Bob to see why.

def•i•ni•tion

Isotopes are atoms of the same element that have different masses. The number of protons and electrons is the same for all isotopes of an element, but the number of neutrons is different, causing each isotope to have a different atomic mass.

On weekends, my Uncle Bob invites his unshaven, lazy friends over to the house to watch football. At any one time, there are somewhere between 4 and 17 "worthless bums" (to quote my aunt) hanging around the house, throwing Chex Mix at the TV. As is often the case, my aunt and cousins will start to fight at some point during the game.

With superhuman speed, Uncle Bob springs into action with his loyal sidekicks. "Shut up, all of you!" they all scream. Whether there are 4 or 17 guests in the house, all of them will work together like a fine-tuned machine to get my aunt and cousins to be quiet.

The same thing works with neutrons in an atom's nucleus. For many atoms, there can be several different numbers of neutrons that serve to stabilize the positive charge in the nucleus. For example, the three protons in lithium can be stabilized by either three or four neutrons. Because these different numbers of neutrons weigh different amounts, the relationships of these two types of atoms to each other are isotopes.

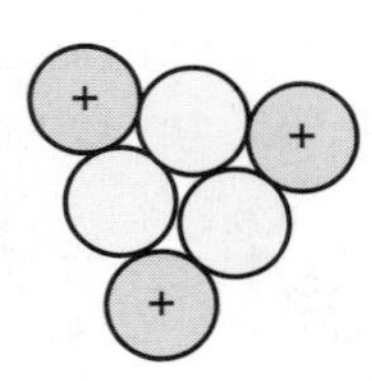

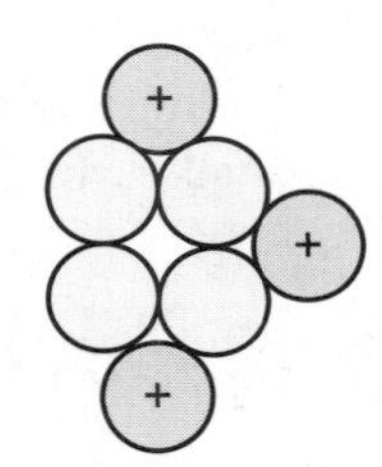

Figure 4.7

Either three or four neutrons can serve to separate the protons in lithium so they don't fly apart.

A, Z, and X Notation

Because all elements have different isotopes, we need a system of symbols to tell them apart from each other. As a result, we have the A, Z, X nomenclature shown below.

$${}^{A}_{Z}X$$

In this system, Z stands for the *atomic number* (the number of protons in the atom), A stands for the *atomic mass* (the number of protons plus the number of neutrons), and X stands for the atomic symbol on the periodic table that denotes each element.

Frequently, this notation leaves out the Z term, because the *atomic symbol* on the periodic table is sufficient for identifying the number of protons. For example, if I were to write ^{35}Cl, you would know that the unseen Z value was equal to 17 because chlorine always has seventeen protons.

Though it's a lot of fun to try to tell people what the A, Z, and X values are for each element, we usually abbreviate the names of the isotopes by saying the name of the element, followed by its atomic mass. As a result, ^{12}C is usually just referred to as "carbon-12."

def•i•ni•tion

The **atomic number** is the number of protons in an element, the **atomic mass** is equal to the sum of the number of protons and neutrons, and the **atomic symbol** is the symbol for each element found on the periodic table.

You've Got Problems

Problem 1: How many protons, neutrons, and electrons are in the following elements?

(a) carbon-14

(b) ^{31}P

(c) nitrogen

Finding Average Atomic Masses

So far, we've discussed the fact that isotopes usually have whole number atomic masses. The atomic mass for ^{1}H is 1 amu, the atomic mass for ^{19}F is 19 amu, and so on. Since this is the case, why are the masses on the periodic table listed as odd decimals? Why is copper's atomic mass listed as 63.55 amu and iron's atomic mass is 55.85 amu?

def•i•ni•tion

The **average atomic mass** of an element is the weighted average of the masses of all of its isotopes.

There hasn't been a mistake. Atomic mass most properly refers to the atomic mass of a particular isotope. The term *average atomic mass* refers to the weighted average of the masses of all the isotopes of an element. Since all elements have more than one isotope, the average atomic masses on the periodic table are all listed as decimals.

The average atomic mass of an element is determined by the following equation:

$$\text{Average atomic mass} = \left(\begin{matrix}\text{mass of}\\ \text{isotope 1}\end{matrix}\right)\left(\begin{matrix}\text{abundance of}\\ \text{isotope 1}\end{matrix}\right) + \left(\begin{matrix}\text{mass of}\\ \text{isotope 2}\end{matrix}\right)\left(\begin{matrix}\text{abundance of}\\ \text{isotope 2}\end{matrix}\right) + \ldots$$

Now let's consider a question. What's the average atomic mass of lithium? Naturally occurring lithium contains two isotopes in the following abundances:

Isotope	Isotopic Mass (amu)	Abundance (%)
^{6}Li	6.015	7.5
^{7}Li	7.016	92.5

Chemistrivia

In the lithium example, why aren't the masses of each lithium isotope listed as exactly 6.000 and 7.000 amu? There are two reasons for this. The first is that protons and neutrons both weigh slightly more than 1.000 amu, throwing off the weight. The second is that some of the mass of protons and neutrons is converted to energy in the nucleus (using Einstein's $E=mc^2$ equation), bringing down the masses of each isotope. Overall, these effects almost cancel each other out, so isotopic masses are always near a whole number of amu. However, except for carbon-12, they are not exactly a whole number.

To solve this problem, we'll plug the masses and abundances of each isotope into the equation for average atomic mass, giving us

$$\begin{aligned}\text{average atomic mass} &= (6.015\text{ amu})(0.075) + (7.016\text{ amu})(0.925)\\ &= 0.45\text{ amu} + 6.49\text{ amu}\\ &= 6.94\text{ amu}\end{aligned}$$

Bad Reactions

When inserting the isotopic abundances into the equation for average atomic mass, make sure to convert the percents into decimals by dividing by 100. Otherwise, you'll find that your average atomic masses are very, very large!

You've Got Problems

Problem 2: Find the average atomic mass of boron using the table below:

Isotope	Isotopic Mass (amu)	Abundance (%)
^{10}B	10.013	19.9
^{11}B	11.009	80.1

The Least You Need to Know

- Atoms are the smallest particle of an element that retains its properties.
- Throughout the history of chemistry, our definition of an atom has changed drastically.
- Atoms contain protons and neutrons in the nucleus, and electrons in orbitals outside the nucleus.
- Isotopes exist because varying numbers of neutrons can be used to stabilize the protons in the nucleus.
- Isotopes are referred to by designations using the symbols A, Z, X.
- The average atomic mass you see on the periodic table is a weighted average of the atomic masses of each of an element's isotopes.

Chapter 5

The Modern Atom

In This Chapter

- The Bohr Planetary Model of the atom
- What spectroscopy is and how it's used
- How quantum mechanics improved on the Bohr Model
- How to write electron configurations and orbital filling diagrams

In Chapter 4, we saw how many of the great minds in science struggled to understand the atom. We also saw how those great minds came up with fairly good (but ultimately wrong) models of the atom. If these guys were so smart, why did they keep screwing up?

As we'll see in this chapter, one of the problems early atomic theoreticians had was that they thought of the atom in terms that they could understand. For example, the Greeks thought the behavior of the universe could be predicted by patterns such as geometric shapes, so they came up with a model of the atom that included lots of neat shapes. Though John Dalton did a good job of refining this idea somewhat, his theories really never varied much from the Greek idea of the perfect atom. J. J. Thomson liked his desserts and came up with a model of the atom that allowed him to dream of feasts to come. These models were intuitive to the people who came up with them based on the information they had at the time.

However, the reason these models didn't work out so well was that they assumed that very small particles (like atoms) must behave in the same way as much larger objects. At the beginning of the twentieth century, people began to realize that this wasn't necessarily the case. As a result, some very unusual models of the atom were devised.

Bohr: What Is It Good For?

While many theorists were busy playing with cathode ray tubes or shooting alpha particles at gold foil, Niels Bohr was busy thinking about hydrogen. As it turns out, when energy is added to a sample of hydrogen it creates a very unusual emission spectrum, as shown in the following figure.

Figure 5.1

When energy is added to hydrogen, light is given off only at particular energies. Bohr used the energies of these bands of light to devise his planetary model of the atom.

Let's go back a second. You've probably seen at some point in your life how a prism can be used to break white light up into a rainbow of colors. This rainbow is referred to as a continuous spectrum, because it breaks light into a continuous band of colors. This differs in appearance from a line spectrum, which shows only certain colors of light that correspond to certain energies.

> **The Mole Says**
>
> The energy of light is closely related to its color. High energy light appears purple, low energy light appears red, and intermediate energies of light have intermediate colors such as blue, green, yellow, and orange.

Bohr was puzzled by the lines given off by hydrogen. What could be causing them? Why didn't hydrogen give off a continuous spectrum? After initially playing with and discarding the idea that demons were responsible, he developed what's now frequently referred to as the planetary model of the atom.

In his planetary model, Bohr suggested that electrons travel around the nucleus in circular orbits, just as the planets revolve around the sun. Because the nucleus is positively charged and the electrons have a negative charge, the pull of the nucleus serves to keep the electrons near it, much as the pull of gravity keeps the planets near the sun. The paths that the electrons traveled were referred to as *orbitals*.

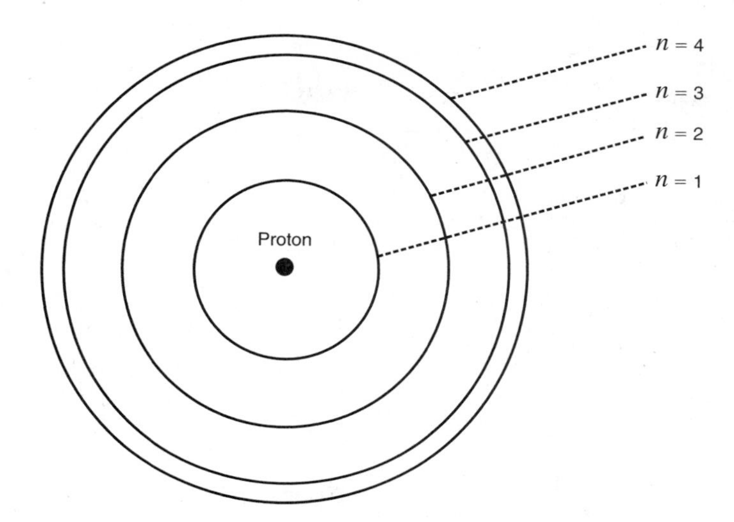

Figure 5.2

Bohr believed that electrons rotated around the nucleus in exactly the same way that the planets revolve around the sun.

That wasn't all. In addition, Bohr believed that the farther away from the nucleus the electrons were, the more energy they had. According to his model, the electrons couldn't have just any random energy—their energies were determined by the order in which the orbitals were located from the nucleus. This order was denoted by the variable n, with the nearest orbital to the nucleus being equal to $n = 1$, the second nearest being equal to $n = 2$, and so on.

In Bohr's model, when energy is added to an atom it causes electrons to move from an orbital close to the nucleus to orbitals that are farther away. The starting orbital is called the *ground state* and the orbital it ends up in is called the *excited state*. However, much like a child who's eaten 15 pounds of chocolate on Halloween, eventually the electrons come down from their excited state and re-enter the ground state.

def•i•ni•tion

The **ground state** is the low-energy state where you can initially find electrons. When energy is added to the atom, the electrons may use this energy to jump into a higher-energy state called the **excited state.**

Now the energy that excited the electrons has to go *somewhere*. After all, it can't just vanish into thin air. When the electron falls back down to the ground state, the energy is given off as light. The energy of the light is equal to the difference in energy between the ground state and the excited state. Because there are several different excited states that electrons can jump into, there are several different energies of light emitted during this process. Bohr was a very happy guy when he saw that the energies of light predicted by his equation matched the energies of light given off in the hydrogen spectrum. He was even happier when this discovery won him the 1922 Nobel Prize for Physics.

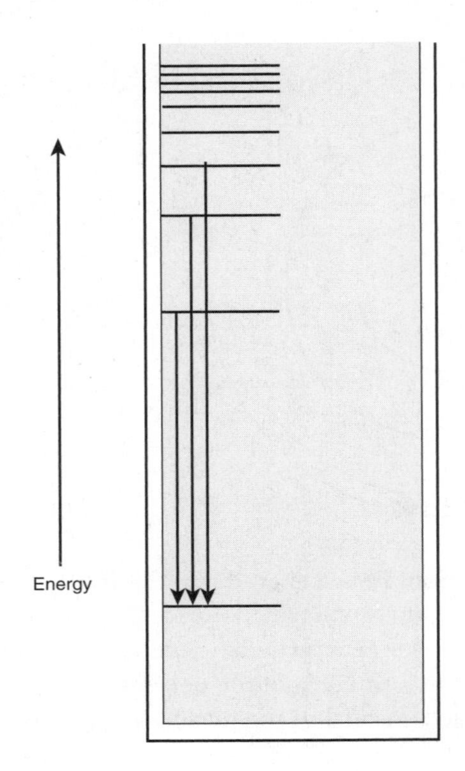

Figure 5.3

The energy of the light that is emitted from an atom is equal to the difference in energy between the excited state and the ground state. As Bohr predicted, the colors of light indicated by his theory matched those in the line spectrum of hydrogen.

Spectroscopy: Reading Between the Lines

Though Bohr spent his time thinking about the emission lines in hydrogen's spectrum, hydrogen isn't the only element that produces a line spectrum. In fact, all elements produce a unique line spectrum, because all elements have unique orbital energies. You can use these emission spectra to identify the elements in a sample of an unknown compound. The method of identifying substances by their spectra is called *spectroscopy*.

def•i•ni•tion

Spectroscopy is a method of identifying unknown substances from their spectra. Because all materials have unique spectra, you can think of these spectra as being "molecular fingerprints."

Although there are many different types of spectroscopy, they all work under the same principle. Some important varieties of spectroscopy include infrared (IR) spectroscopy, UV-vis spectroscopy (which uses light in the ultraviolet and visible range of the spectrum), and nuclear magnetic resonance spectroscopy (a.k.a. NMR spectroscopy, which uses pulsed radio waves).

A Quantum Leap into Quantum Mechanics

Unfortunately for Bohr, his model didn't properly explain how atoms behave. Fortunately, quantum mechanics came to the rescue!

What Do Orbitals Look Like?

The Bohr model was definitely a step in the right direction. After all, if the model could be used to predict the orbital energies of hydrogen, there must be something to it! Unfortunately, there was one small problem: it was ineffective at predicting the orbital energies of any other element. As a result, the hunt to come up with something more effective was on.

After a whole lot of work by a whole lot of really famous guys you've probably never heard of, a new model of the atom was born. This new model, called quantum mechanics, sums up our current understanding of how atoms work.

The problem with Bohr's model was that there weren't enough variables in his equation to do a good job of predicting the orbital energies of elements other than hydrogen. Eventually, the Schrödinger equation was written to explain these orbital energies. Because the Schrödinger equation is pretty complicated, we'll save that for another book.

When the Schrödinger equation was put together, a really strange thing happened. Whereas previous models of the atom had predicted that electrons were particles that zoomed around in predictable paths, the Schrödinger equation predicted that it was possible to predict where electrons would probably be found, but impossible to predict exactly where you could find them. However, you can express the probability of finding an electron with a mathematical function called a wavefunction. As a result, orbitals went from being two-dimensional circular orbits to 3-D shapes, some of them very weird.

Chemistrivia

The idea that it's impossible to predict exactly where an electron can be found is referred to as the Heisenberg Uncertainty Principle. According to this principle, it's impossible to simultaneously know both the momentum and the position of any object. You can know either where an object is or where it's going, but not both at the same time. The reason you've never noticed this while throwing a baseball around is that this limitation is small enough that it can be ignored for large objects.

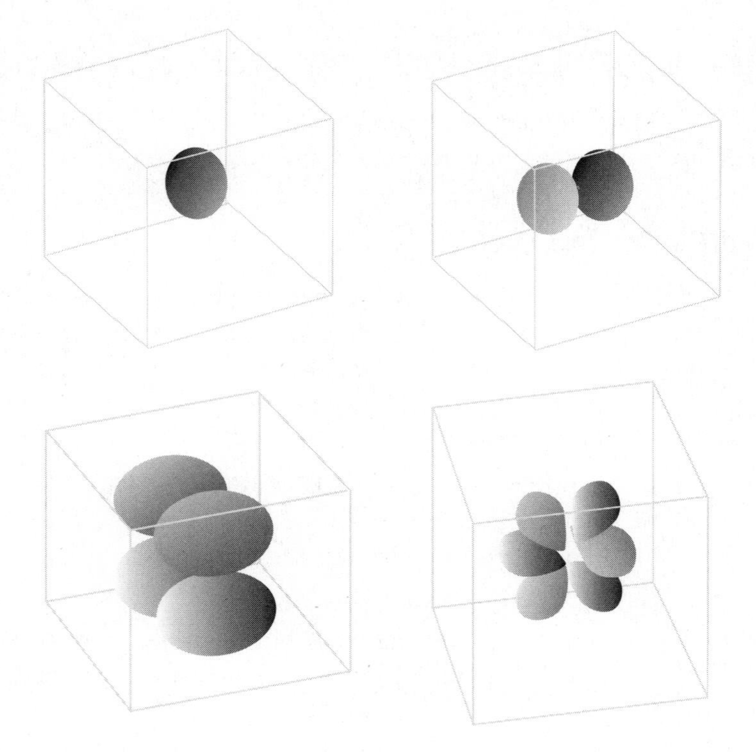

Figure 5.4

These diagrams represent the four types of atomic orbitals. More about them later.

You may be wondering how somebody just learning about quantum mechanics can understand how wavefunctions work. Since differential equations are really hard, learning the Schrödinger equation directly isn't practical for most of us. Instead, we'll get a little help from our friendly neighborhood garden sprinkler.

Imagine that you've just turned on a sprinkler and are watering your lawn. Most of the water that leaves the sprinkler head lands pretty close to the sprinkler itself, and as you move away from the sprinkler, fewer and fewer drops hit any one location. If you wanted to measure exactly how much water lands at any location, you could set out little rain gauges all over your lawn. Let's make a chart that shows what these rain gauges would look like.

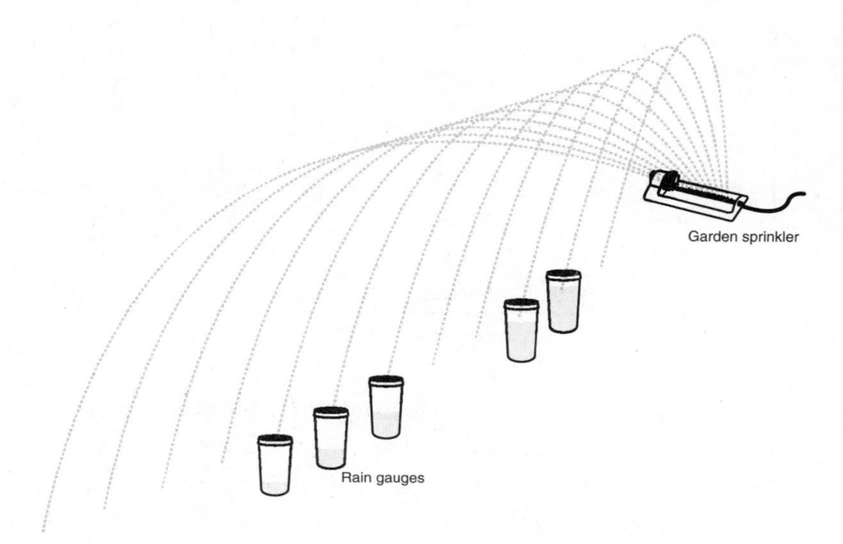

Figure 5.5

By placing rain gauges near a lawn sprinkler, we get an idea of where future raindrops might land. This type of chart is referred to as a "probability distribution."

Now for a couple of difficult questions:

1. What is the maximum distance that a drop can land from the sprinkler? This is a tough one. There isn't any limit to how far a drop can go—in fact, if you've ever stood near a sprinkler, you can see that a few stray water drops always land outside the area that really gets a good dousing. So, what's the maximum distance? There isn't one!

2. If you were to inject a single drop of colored water into the hose, where would it land? From the accompanying figure we can see that it will probably land near the sprinkler, but there's also a chance that it will land farther away. Unfortunately, our diagram doesn't tell us for sure where the drop will land, though it does tell us the probability of finding a drop at any one location.

The garden sprinkler figure is referred to as a "probability distribution" because it gives us an idea of where we're likely to see the water drops fall. In the same way, we can draw a probability distribution that gives us an idea of where we're likely to find an electron in an atom. Just like the sprinkler example, we can't know from this chart where an electron will be. It does, however, give us some idea as to where we'll find it.

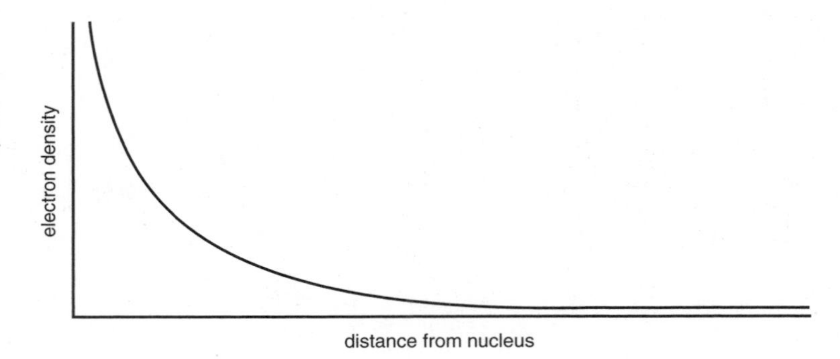

Figure 5.6

From this diagram, you can see that electrons can be compared to the individual water drops from a lawn sprinkler. You can never tell exactly where they'll land, but the probability distribution tells you where they're most likely to be found.

If we sketch these probabilities out on paper, we can see that s-orbitals have a spherical shape, with the electrons most likely to be found near the center and less likely to be found near the edges. The sketch shown in the following figure is what an s-orbital looks like.

Figure 5.7

Though s-orbitals are spherical with a fuzz of decreasing probability near the edges, they're usually just drawn as spheres to make them easier to visualize.

Where Do These Fancy Orbital Shapes Come From?

A few pages ago we mentioned that the equations that describe electrons are a lot fancier in quantum mechanics than they are in the Bohr model. To make these equations work, four variables called quantum numbers are needed, rather than the one variable (n) that Bohr used.

- The principal quantum number, n. This is the same n that Bohr used, and is used to describe the energy level of an electron. The allowed values for n are 1, 2, 3, … to infinity.

- The angular momentum quantum number, l. This quantum number is used to determine the shape and type of the orbital. Possible values for l are 0, 1, 2, and so on up to $(n-1)$. For example, if $n = 2$, the possible values for l are 0 and 1. A spherical s-orbital is defined as $l = 0$, a dumbbell-shaped p-orbital is defined as $l = 1$, an oddly shaped d-orbital is defined as $l = 2$, and a really weird f-orbital is defined as $l = 3$.

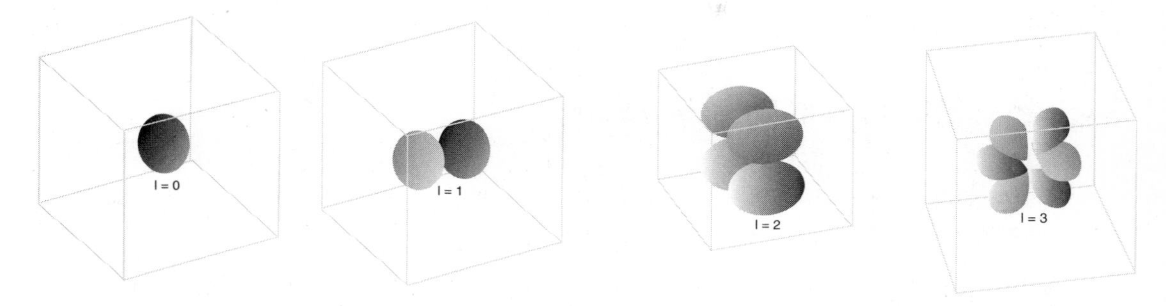

Figure 5.8

The value of l determines the shape and type of orbital being described.

- The magnetic quantum number, m_l. The magnetic quantum number determines the direction that the orbital points in space. Possible values for m_l are all the integers from $-l$ through l. For example, if $l = 2$, the possible values of m_l are –2, –1, 0, 1, 2. Because p-orbitals are denoted by $l = 1$, there are three possible directions that they can point, denoted by $m_l = -1, 0, 1$. These three directions lie along the x-, y-, and z-axes (see accompanying figure).

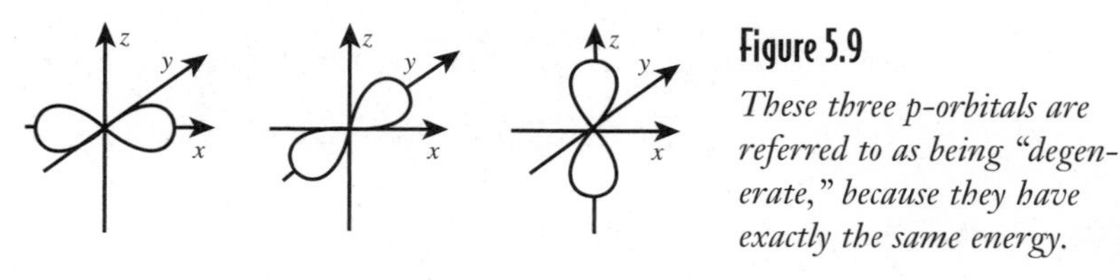

Figure 5.9

These three p-orbitals are referred to as being "degenerate," because they have exactly the same energy.

Similarly, there is one s-orbital per energy level—since $l = 0$, the only value for $m_l = 0$. There are also five d-orbitals per energy level ($l = 2$, so $m_l = -2, -1, 0, 1, 2$) and seven f-orbitals ($l = 3$, so $m_l = -3, -2, -1, 0, 1, 2, 3$).

- The spin quantum number, m_s. Possible values for m_s are +½ and –½. The reason we need a spin quantum number comes from the *Pauli exclusion principle*, which states that no two electrons in an atom can have the same set of quantum numbers. Now, if we only had the

def•i•ni•tion

The **Pauli exclusion principle** states that no two electrons in an atom can have the same four quantum numbers.

You've Got Problems

Problem 1: What type of orbital is described by an electron with the quantum numbers $n = 3$, $l = 2$, $m_l = 1$, $m_s = +½$?

quantum numbers n, l, and m_l, the Pauli exclusion principle would force each electron to be in its own orbital. Because it's well known that orbitals are capable of holding two electrons, we need to define a fourth quantum number that allows us to distinguish between them.

Electron Configurations (the Long Way)

So far, we've talked about the nature of orbitals with the quantum model and the variables that define their shapes. What we haven't yet discussed is where the electrons for particular elements can be found.

Before we can understand how electrons occupy the orbitals in an atom, we need to learn which orbitals have the lowest energies and which have the highest energies. After all, electrons will fill up low-energy orbitals before they start to fill high energy orbitals, so if we can order the orbitals by increasing energy, we can determine where all of the electrons in an atom go.

In each principal energy level, s-orbitals have the lowest energies, followed by p-orbitals, d-orbitals, and f-orbitals (which have the highest energies). As a result, electrons will go to s-orbitals before entering p-orbitals and so on for d- and f-orbitals.

Once we know where all the electrons in an atom belong, we can write electron configurations for the atom. An electron configuration is nothing more than a list of orbitals that contain the electrons in an atom. Electron configurations usually contain many terms, each of which have the general format:

$$n\ (\text{type of orbital})^{\text{number of electrons in that type of orbital}}$$

Memorizing the electron configurations of each element is possible, but not much fun. Instead, we'll use the periodic table to help us with our task. To make life really easy, we'll label our periodic table in a way that gives us the most possible useful information about electron configurations.

Let's look at some examples:

- Hydrogen has only one electron. As you can see from the periodic table in the tear-out card inside the front cover of this book, hydrogen is in the first row of the periodic table, making the principal quantum number 1. The type of orbital the ground state electron lies in (shown by the color of the element on this handy card) is an s-orbital. As a result, the electron configuration for hydrogen is said to be $1s^1$, where the one in front stands for the principal quantum number and the one after the "s" stands for the number of electrons in that orbital.

- Helium has two electrons. Since helium is also in the first row of the periodic table, n is still equal to 1. Helium is in the s-region of the periodic table, meaning that the electrons are in the 1s orbital. As a result, helium's electron configuration is $1s^2$.
- Lithium has three electrons. The first two electrons are the same as those in helium. So the first term will be $1s^2$. The third electron is represented by lithium's position on the periodic table, which is in the second energy level and in the s-region of the periodic table, making the second term $2s^1$. Putting them together, the electron configuration is said to be $1s^22s^1$.

Chemistrivia

You've probably noticed that electron configurations are very repetitive. This is because of the Aufbau principle, which states that every element has the same number and placement of electrons as the one before it, plus one extra. As a result, once you can do simple electron configurations, harder ones just fall into place!

- Beryllium has four electrons. The first two are the same as helium (making the first term $1s^2$) and the second two are in the 2s orbital (making the second term $2s^2$). Overall, the electron configuration is $1s^22s^2$.
- Boron has five electrons. The first two are the same as those in helium (making the first term $1s^2$) and the second two are the same as beryllium (making the second term $2s^2$). The fifth electron is represented by boron's position on the periodic table, which is in the second row and p-section of the table. The third term, then, is $2p^1$, making the overall electron configuration $1s^22s^22p^1$.
- In the same way, the electron configuration for carbon's six electrons is $1s^22s^22p^2$, and the electron configuration for nitrogen's seven electrons is $1s^22s^22p^3$.

There are only a few variations from this pattern that we can see. Let's use scandium (Sc) as an example. You might think that the electron configuration would be $1s^22s^22p^63s^23p^64s^24d^1$, which isn't a bad guess based on what we've seen earlier. However, something very important to remember is that the number in front of the d-orbitals is always one behind the energy level of the s-orbitals. As a result, it's not a 4d orbital, it's a 3d orbital. The correct electron configuration, then, is $1s^22s^22p^63s^23p^64s^23d^1$.

You've Got Problems

Problem 2: Write the electron configurations for the following elements:

(a) gallium (Ga)

(b) iridium (Ir)

In the same vein, the f-orbitals also cause some trouble when doing electron configurations. Instead of having the same *n* value as the s-orbitals, f-orbitals are two behind. As a result, cerium (Ce) has an electron in the 4f orbitals, not the 6f orbitals.

For those of you who want to see how the whole thing plays out, we'll use the example of lead (Pb). The complete electron configuration for lead is $1s^2 2s^2 2p^6 3s^2 3p^6 4s^2 3d^{10} 4p^6 5s^2 4d^{10} 5p^6 6s^2 4f^{14} 5d^{10} 6p^2$.

Electron Configurations (the Short Way)

Now that you know how to write electron configurations the long way, we can concentrate on doing it in a much less tedious fashion. For those of you who skipped the long version to see what the short way was, go back and learn the long way or you won't have any idea what we're talking about in this section.

As you've undoubtedly been able to tell, electron configurations involve a lot of repetition. Every element from helium on has the "$1s^2$" term. As a result, it's usually not handy to write out every single term for every element. Instead, what we'll do is write only the last few terms for each element, starting instead at the previous noble gas (they're the elements in the far right column of the periodic table—we discuss them in Chapter 7). To denote that you've done this, write the symbol of the noble gas in [brackets].

You've Got Problems

Problem 3: Write the abbreviated electron configurations for the following elements:

(a) yttrium (Y)

(b) polonium (Po)

Example: The electron configuration of phosphorus (P) can be written in shorthand as [Ne] $3s^2 3p^3$. Isn't that nicer than writing out all those other terms? Similarly, you can abbreviate something as obnoxious as plutonium (Pu) as [Rn] $7s^2 5f^6$, a great improvement over the old method.

Orbital Filling Diagrams and Hund's Rule

When I get on the bus I don't like to sit next to other people if I can avoid it. Though somebody may look okay when I first sit next to them, invariably they start picking their nose or sneezing on me or eating a meatball sandwich as soon as I get comfortable. Over the years, I've learned that I'd better head for empty seats if it's at all possible. I've noticed that I'm not the only person to act this way—most people do the same thing.

Electrons work in the same way. Let's consider the case of carbon, which has the electron configuration $1s^2 2s^2 2p^2$. Since there is only one s-orbital per energy level, the electrons in the 1s and 2s orbitals are stuck pairing up. However, since there are three p-orbitals per energy level, the two electrons in the 2p orbitals will choose to go into different p-orbitals rather than sticking around together. If we were to sketch this, with low-energy orbitals being further down on the diagram than higher-energy orbitals, we'd see the following.

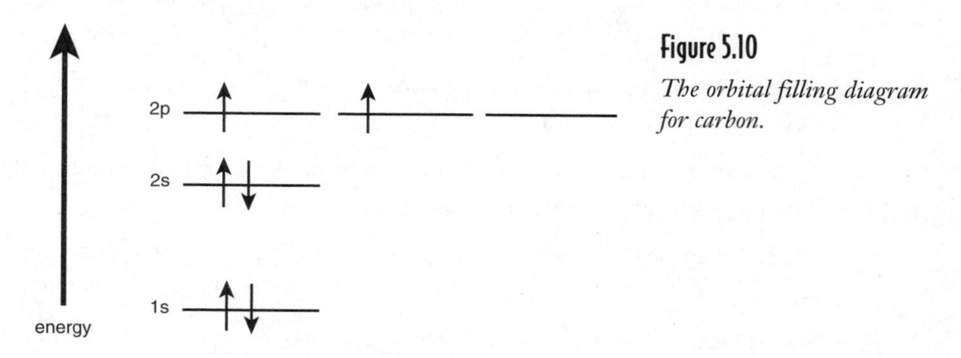

Figure 5.10

The orbital filling diagram for carbon.

The fact that electrons will want to stay unpaired whenever possible in orbitals with equal energies is called *Hund's rule.* Hund's rule states that electrons will stay unpaired whenever possible in orbitals with equal energies.

Of course, if we add enough electrons, we get into a situation where we simply have to start putting two electrons together in the same orbital. An example of this can be found with oxygen, which has the electron configuration $1s^2 2s^2 2p^4$.

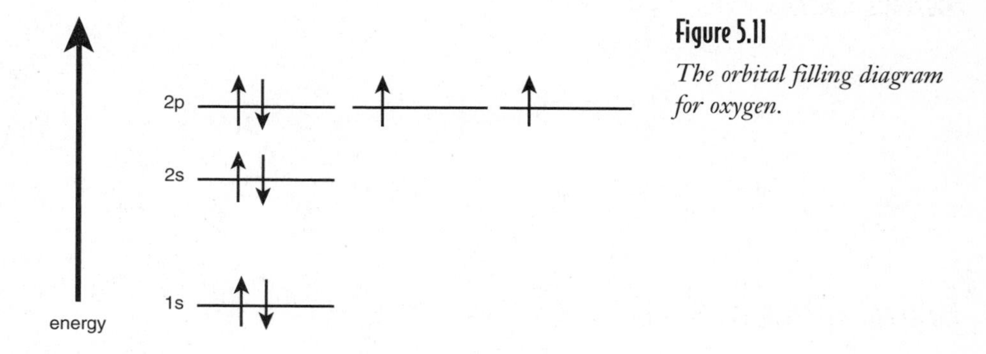

Figure 5.11

The orbital filling diagram for oxygen.

As you can see, the four electrons in the 2p orbital have been forced to pair up. However, since we have four electrons and three orbitals, only one of the orbitals needs to contain paired-up electrons. We see the same situation happening with other elements as well.

You've Got Problems

Problem 4: Draw the orbital filling diagram for chlorine.

The Least You Need to Know

- The Bohr model of the atom did a good job of explaining why line spectra are formed but could only predict the orbital energies of hydrogen.
- Elements can be identified by examining their line spectra using a process called spectroscopy.
- Quantum mechanics improved on the Bohr model by introducing three more quantum numbers to describe electrons.
- Electron positions cannot be determined with perfect accuracy. The best you can do is draw graphs showing where they are most likely to be found.
- Electron configurations are used to describe the locations of each of the electrons in an atom.
- Hund's rule states that electrons prefer to be unpaired in orbitals with equal energies.

Part 2: A Matter of Organization

In this part, I'm going to teach you how to cheat your way through chemistry. No, I'm not going to identify parts of your body that are particularly good at absorbing ink. Instead, we'll learn tricks that can help us figure out what chemicals are likely to do.

We'll start by learning about elements, compounds, and mixtures. I'll then tell you all about the best cheat sheet ever: the periodic table. Unlike cheat sheets that are really small and need to be hidden from teachers, the periodic table can be put on your desk in plain sight, and there's nothing your teacher can do to keep you from using it!

After we've mastered the periodic table, we'll learn about how ionic and covalent compounds work. Finally, we'll end up with the concept of the mole—don't worry, this one is easy, too!

Chapter 6

Elements, Compounds, and Mixtures

In This Chapter

- Elements
- Compounds
- Homogeneous mixtures
- Heterogeneous mixtures
- Colloids
- Methods commonly used to separate mixtures

Let's review what we've learned about chemistry. In the previous five chapters, we've managed to learn … what a single atom looks like. This is nothing to sneeze at, but clearly we still have quite a way to go before we can call ourselves master chemists.

In this chapter, we're finally going to investigate the mysteries of elements, compounds, and mixtures. When I tell my students we're going to learn about these in class, they usually answer with groans and sighs, saying, "We already know about this stuff." As with the atom, they're sometimes surprised to find that there's more here than meets the eye!

Pure Substances

If you're a salesperson, you know the word "pure" is a great way to sell something. When consumers think of the word "pure," they imagine something that's perfect in every possible way. Have cockroaches been crawling in the potato salad? Of course not—it's pure!

To a chemist, the word pure has a somewhat different connotation. When we say something is pure, we mean that only one substance is present in the material, and that it is completely uniform in composition. By uniform composition, we mean that if you take a sample from one part of a material and compare it with a sample from another part of the material, both samples will be identical in every way. Things that have a uniform composition are said to be "homogeneous."

Pure Substance #1: Elements

Before you read any further, I want you to come up with a definition of the word *element*. Since you probably already know what an element is, this should be a really easy task.

Not so easy, huh? Just like with the word "atom," "element" is a surprisingly difficult term to define. You probably have an internal idea of what an element is, and you probably already know that a list of all the elements can be found on the periodic table, but putting it in words is often difficult.

def•i•ni•tion

An **element** is a substance that cannot be chemically decomposed into simpler substances. The periodic table is a list of all the known elements.

For chemical purposes, an element can be defined as a substance that cannot be chemically decomposed into simpler substances. You may already be aware that nuclear reactions are able to break elements apart into even simpler particles, but these particles are unimportant for most chemical purposes.

Pure Substance #2: Compounds

The other type of pure substance is a chemical compound. *Compounds* are pure substances made up of two or more elements in defined proportions. Unlike elements, compounds can be broken into simpler parts using chemical reactions. These pieces, of course, are just the elements that make up the compounds. For example, it is possible to convert sodium chloride back into pure sodium and pure chlorine using a process called electrolysis.

You can typically figure out whether a material is an element or a chemical compound by looking at its name or formula. The names of many chemical compounds contain two words (e.g., "sodium chloride," "magnesium sulfate") and the symbols of chemical compounds contain more than one atomic symbol (e.g., "NaCl," "$MgSO_4$").

def•i•ni•tion

Compounds are pure substances made up of two or more elements in defined proportions. By using chemical reactions, compounds can be broken into their constituent elements, though this is usually difficult.

Shake It Up: Mixtures

Mixtures are materials that contain more than one type of pure element or compound. For example, even if cockroaches haven't been crawling around on it, potato salad is referred to as being a mixture because the potatoes are made of different chemical compounds than the mayonnaise. Likewise, salt water is a mixture because it consists of pure sodium chloride ("salt") mixed with pure water.

The components in a mixture may or may not exist within the same phase. For example, if we are talking about a mixture of two solids, each solid may be distinctly different from one another, such as you'd find in a bowl of gumdrops where each gumdrop is different from the others. In a mixture of two liquids, there may or may not be more than one phase—generally, liquids tend to dissolve in one another, and it's uncommon to find mixtures of more than two or three liquids that don't dissolve one another, called multiphase liquid mixtures. An example of a multiphase liquid mixture would be oil and vinegar. Mixtures of gases are always well mixed and don't readily separate into separate phases. An example of a gaseous mixture would be air, in which the oxygen, nitrogen, and other stuff don't separate into separate components.

Homogeneous Mixtures

We've already seen that elements and compounds are homogeneous materials because they have a completely uniform composition. Some mixtures can also be said to be homogeneous because they contain two or more pure substances mixed together in a uniform fashion. These mixtures are called, straightforwardly enough, *homogeneous mixtures*.

def•i•ni•tion

When two pure substances are mixed evenly, the resulting material is referred to as a **homogeneous mixture,** also known as a solution.

You can usually tell if something is a homogeneous mixture by looking at it. If something contains more than one pure material and appears uniform to the naked eye, it's probably a homogeneous mixture. Examples of homogeneous mixtures include salt water, air, and stainless steel.

Heterogeneous Mixtures

Most of the things that we encounter in our everyday lives aren't evenly mixed. For example, I had a jelly-filled doughnut with powdered sugar on it for breakfast this morning. When I bit into it, the powdered sugar went all over my face and the jelly plopped into my lap. Because the material separated into several distinct parts, the doughnut could be said to be a *heterogeneous mixture*.

def•i•ni•tion

Mixtures are said to be **heterogeneous** if they contain two or more unevenly mixed parts.

Heterogeneous mixtures are pretty easy to spot because they clearly contain several different components. Examples of heterogeneous mixtures include my shoe, that annoying dirt-eating kid down the street, and the 1987 Boston Red Sox.

You've Got Problems

Problem 1: Identify each of the following as being either a homogeneous or heterogeneous mixture:

(a) Turkey stuffing

(b) Sugar water

(c) Chunky peanut butter

Colloids: Somewhere in Between

Earlier, I said that something was *probably* a homogeneous mixture if it had more than one component and appeared uniform to the naked eye. However, let's imagine that you make a watercolor paint by combining paint powder with water. When the two have mixed, the resulting paint consists of fine pigmented particles floating around in water. Because these paint particles are so small, they never settle at the bottom of the container.

So what's the deal with this stuff? It's hard to say that it's a solution because it has both solid and liquid components that can be separated from each other by filtering. On the other hand, if you were to look at paint under a microscope, it would still look completely uniform. As a result, this material is somewhere on the borderline between a solution and a heterogeneous mixture. Such materials are referred to as *colloids*.

def•i•ni•tion

Colloids are materials in which one type of particle is suspended in another without actually having been dissolved.

There are many types of colloids, categorized by the phases of matter they contain. The main types are listed below:

- Aerosols consist of liquid or solid particles suspended in a gas. Examples of aerosols include smoke and fog.
- Foams are formed when a gas is suspended in a liquid or a solid. Common examples are shaving cream and marshmallows.
- Emulsions are formed when particles of a liquid are suspended in another liquid or a solid. Mayonnaise and butter are common examples of emulsions.
- Sols are made when solid particles are suspended in a liquid. Examples include paints, gelatin, and blood.

Chemistrivia

One easy way to distinguish colloids from solutions is to shine a light through them. Because the molecules in a solution are very small (usually single molecules), they don't reflect light. Colloids, on the other hand, have particles large enough to reflect light. As a result, light beams are visible within colloids but not within solutions. This effect is referred to as the Tyndall effect. In some cases, this effect is so pronounced that the milkiness of the liquid makes it impossible to see through it at all.

Separating Mixtures

One of the most important things that chemists do is separate mixtures. For example, when I was working in a chemistry lab and I would go out for lunch, it was of vital interest to me that I effectively separate the pickles from the rest of the hamburger, lest I become sick later in the day. Other chemists I know have also found the ability to separate mixtures to be important when eating trail mix or scraping mud from their shoes. Generally, I've found it's easier to separate the components in a heterogeneous mixture (e.g., the pickles from hamburgers) than the components in a homogeneous mixture (e.g., the rum from a Piña Colada) because it's easier to pick things apart when you can see the different components.

It's good that chemists get so much practice separating mixtures like these in their everyday lives because mixture separation is important for other purposes as well. Let's take a look.

Filtration

One of the simplest methods used to separate mixtures is filtration. If one of the components is a liquid and the other is a solid, filtration is as easy as pouring the whole mixture through filter paper. An everyday example of filtration can be seen in a coffee maker, where the coffee passes through a paper filter but the grounds do not.

Distillation

When one compound is dissolved in another, or when two liquids are mixed together, the most commonly used method to separate them is distillation. In a distillation, the mixture is slowly heated over a Bunsen burner or hotplate. Because the components in a mixture have different boiling points, one of them will boil before the other. The vapor from this compound can be collected from a condenser, enabling it to be isolated in a pure form. A distillation apparatus is shown in the following figure.

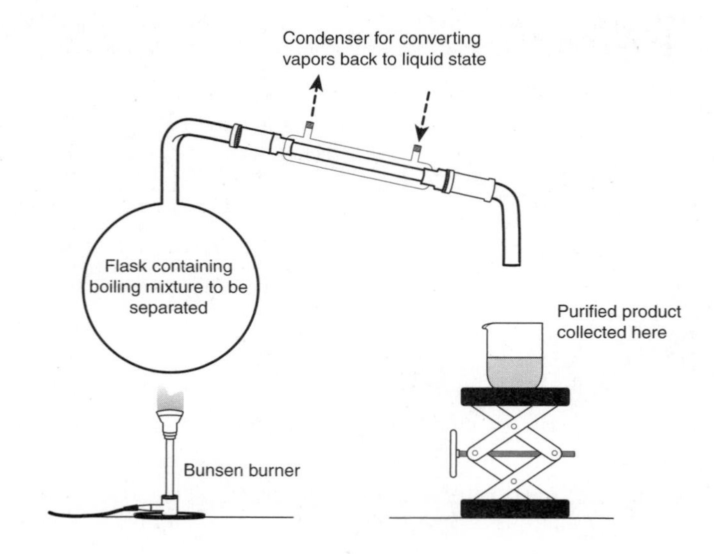

Figure 6.1

A distillation apparatus (frequently referred to as a "still") uses the different boiling points of the components in a mixture to separate them.

Chromatography

At some time or another, all of us have marked our shirt with a pen. Sometimes we get lucky and the ink doesn't stick well to the fabric—in these cases, we can clean the shirt by putting it in the wash. Sometimes we get very unlucky and the ink sticks to the fabric so well that it's there for good, no matter how many times it's washed with bleach and detergent.

In the same way, chemical substances can frequently be separated from one another based on how well they stick to a solid. The use of this difference in "stickiness" to separate the components of a mixture is referred to as chromatography.

Typically, chromatography is performed by placing a mixture of two or more chemicals into a glass column filled with silica. When an organic solvent such as ethyl acetate or alcohol is poured through the column, one of the components of the mixture will tend to stick to the silica better than the other. As a result, the less sticky one will pass through the column more quickly, while the stickier one will take a little longer.

Extraction

Let's say that you have a compound dissolved in a liquid that you want to remove. For example, you have a small amount of salt dissolved in oil and want to remove it. How would you do this?

Though distillation could do the job, it takes a long time and considerable effort. An alternate way of making this separation is to find a liquid that isn't soluble with the first liquid and that's better at dissolving the salt than the oil is. When the two liquids are mixed and shaken, the salt will tend to move from the oil (where it's not very soluble) into the water (where it is). When this process is complete, it's a simple matter to pour out the water, leaving behind the pure oil.

The Least You Need to Know

- Elements and compounds are both examples of pure substances.
- Mixtures are formed when two or more pure substances are combined. If their composition is uniform, the mixture is referred to as a homogeneous mixture or solution. If the composition is not uniform, it's referred to as a heterogeneous mixture.
- Colloids are materials that are on the dividing line between heterogeneous mixtures and solutions.
- The most common methods used to separate mixtures are filtration, distillation, chromatography, and extraction.

Chapter 7

The Periodic Table

In This Chapter

- The development of the modern periodic table
- The properties of metals, nonmetals, and metalloids
- The main groups of elements in the periodic table
- The octet rule and the shielding effect
- The four main periodic trends

The periodic table is one of the most amazing things ever invented. Whereas most tables just have random lists of information, the periodic table tells you what elements are present, how much they weigh, and how many protons they have. As an added bonus, it places them in an order so you can figure out the properties of the elements and their electron configurations. It's an amazing invention, probably the closest thing to magic that modern science has ever produced.

The reason the periodic table was invented in the first place was so you wouldn't *have* to memorize everything about the elements. In this chapter, I'm not going to give you any hints about how to memorize the periodic table because that's a pointless activity. Instead, we're going to discuss how you can use the periodic table to learn about the elements.

Early Periodic Tables

In 1871, a Russian chemistry professor named Dmitri Mendeleev was hard at work writing a chemistry book for his students. Old Dmitri was having trouble when it came time to discuss elements because the only real approach to learning about them involved memorizing their individual properties. Needless to say, most students didn't enjoy this.

In any case, Dmitri decided that memorizing all of the elements was a real pain and decided to create an organizational method that would make the elements a little easier to understand. After much trial and error (and error and error), he realized that when he placed the elements in order of their atomic masses, he could arrange them into columns of elements with similar properties.

Chemistrivia

The first guy to come up with a periodic table based on atomic mass was the German chemist Julius Lothar Meyer. Unfortunately, when he published his own textbook in 1864, he only included 28 of the known elements in his periodic table. Because Mendeleev's table included all of the known elements, he usually gets credit for inventing it.

When coming up with his periodic table, Mendeleev, like most of us, made some mistakes. Fortunately for him, these mistakes weren't discovered until six years after his death. In 1913, British physicist Henry Mosely discovered the concept of the atomic number, which turned out to be a better method for ordering the elements into a table than ordering them by increasing mass. This discovery was the start of a promising scientific career, which ended two years later when Mosely was killed during World War I.

The Modern Periodic Table

Our modern periodic table is pretty much like Mendeleev's. If you take a look at the tear card in the front of this book, you'll find the most recent version of the periodic table. I strongly recommend pulling it out of the book (if you haven't done so already) so you can refer to it through the rest of this discussion.

The vertical columns in the periodic table are called *groups* or *families* (the terms are synonymous). Elements in the same family have similar chemical and physical properties. This can be explained partly by the fact that elements in the same family have similar electron configurations. For example, the electron configuration of lithium is

[He]2s^1 and the electron configuration of sodium is [Ne]3s^1. Because both of them have one electron in the outermost s-orbital, they have similar properties.

The horizontal rows in the periodic table are called *periods*. Elements in the same period don't have much in common except that the energies of their outermost electrons are similar. We usually find that elements near each other in the same period have more shared properties than distant elements because the electron configurations are more alike.

def•i•ni•tion

Families or **groups** are the columns in the periodic table, and consist of elements with similar chemical and physical properties. **Periods** are the horizontal rows in the periodic table, and consist of elements that have little in common with one another aside from the energies of their outermost electrons.

Your Roadmap to the Periodic Table

Because the periodic table is broken into regions with similar properties, we can use it to give us a general idea of the behavior of unfamiliar elements based on their locations. To do this, we must first learn the characteristics of the various areas of the periodic table.

Metals, Nonmetals, and Metalloids

If you look at the elements on the right side of the periodic table, you see a line that starts in front of boron (B) and ends between polonium (Po) and astatine (At). This line marks the separation between metals and nonmetals on the periodic table—metals are found to the left of the line and nonmetals are found to the right.

To complicate matters, the elements immediately surrounding this line have properties of both metals and nonmetals. As a result, they're referred to as metalloids or semimetals. Boron (B), silicon (Si), germanium (Ge), arsenic (As), antimony (Sb), tellurium (Te), and polonium (Po) are all metalloids.

The main properties of metals, nonmetals, and metalloids are shown in the following table.

Property	Metals	Nonmetals	Metalloids
Luster	Shiny	Not shiny	Not shiny
Hardness	Hard	Brittle	Brittle
Malleable and ductile	Yes	No	No

continues

(continued)

Property	Metals	Nonmetals	Metalloids
Conduct heat and electricity	Yes	No	Semiconductors
State	Solids (and 1 liquid)	Varies	Solid

We'll discuss more about why each type of substance has these properties in Chapter 11.

Periodic Families

As mentioned earlier, elements in the same family in the periodic table have similar properties. Some of the most important families are these.

- Group 18—Noble Gases: noble gases are almost entirely unreactive. This lack of reactivity stems from the fact that completely filled s- and p-orbitals (see Chapter 5) makes them very stable. As a result, very few noble gas compounds can be made. Noble gases are used in advertising signs, toy balloons and blimps, and as inert atmospheres in locations where chemical reactions would be undesirable.

The Mole Says

Because of the high stability of the noble gas electron configuration, other elements gain or lose electrons to obtain the electron configuration of the closest noble gas. This tendency to become like the closest noble gas is referred to as the "octet rule" because most noble gases contain a total of eight s- and p- electrons in their outer shell (helium has only two electrons, as there are no p-orbitals in the first energy level). As we'll see later in this chapter, this tendency to gain or lose electrons is the driving force for many chemical processes.

- Group 1 (except for hydrogen)—Alkali metals: alkali metals are highly reactive, combining readily with air and water. Though they are metallic, their densities are low (only rubidium and cesium are denser than water) and they are soft enough to be cut with a knife. The high reactivity of the alkali metals comes from the fact that they have only one more electron than the very stable noble gases. As a result, they react vigorously in attempts to lose this extra electron. Alkali metals can be found in sodium vapor fog lamps and in the psychiatric drug lithium carbonate.
- Group 2—Alkaline earth metals: the alkaline earth metals have many of the same properties as the alkali metals, although they are less extreme. For example,

most alkaline earth metals react with air and water, but much less violently than the alkali metals. Alkaline earth metals are generally harder than the alkali metals, but are still softer than many other metals. The diminished reactivity of the alkaline earth metals can also be explained by their electron configurations. Because they have to lose two electrons to become like a noble gas, they are somewhat less reactive than the alkali metals. Alkaline earth metals can be found in chalk (calcium carbonate), high-end bicycle frames (beryllium), and in beverage cans (a magnesium/aluminum alloy).

- Groups 3–12—d-Transition Metals (frequently called simply transition metals): though properties of the d-transition elements vary greatly, many of them are hard, have high melting and boiling points, are excellent conductors of heat and electricity, and have moderate to low reactivities. Transition metals are used for a variety of purposes such as structural materials in buildings, power transmission lines, jewelry, and knives.
- f-Transition Metals (sometimes called "inner transition metals"): the f-transition elements consist of the two rows at the bottom of the periodic table, and aren't properly said to be in any of the 18 "groups." The top row, also known as the lanthanides, consists of shiny, reactive metals. Because many lanthanides emit colored light when hit by a beam of electrons, they are used as phosphors in television sets and fluorescent light bulbs. The bottom row, also known as actinides, are primarily radioactive elements that have a wide variety of uses such as nuclear fuel sources, smoke detectors, and atomic bombs.
- Group 17—Halogens: these are highly reactive elements that combine readily with metals to form salts. This extremely high reactivity comes from their electron configurations—because they need only one more electron to have the electron configurations of a noble gas, they react vigorously to pick up that electron whenever possible. The halogens are diatomic elements, meaning that they have the general formula X_2 (for example, fluorine exists as F_2 in its pure form). Fluorine and chlorine are gases under standard conditions, while bromine is a liquid and iodine is a solid. Halogens are widely used in water treatment, in the manufacture of other chemicals, and in plastics such as Teflon.

Bad Reactions

Some students think that diatomic elements always have to have a "2" under their symbols no matter where they show up. Not so! Make sure you know that this is only true for the pure element. As a result, chlorine gas has the formula Cl_2 because it contains only chlorine atoms, while sodium chloride has the formula NaCl (without the "2"!) because chlorine has bonded to sodium.

- Hydrogen—The weirdo: hydrogen has properties unlike any other element in the periodic table. Though it's found near the metallic region of the periodic table, it is a nonmetallic gas. It is diatomic, found as H_2. Hydrogen reacts slowly with other elements at room temperature but may react blindingly fast when heated or catalyzed. Hydrogen is used in the manufacture of ammonia, sulfuric acid, and methanol, and is widely discussed as a fuel alternative to gasoline.

The Mole Says

If you compare the periodic table in your textbook to the tear card in the front of this book, you may find that there's a slight difference in placement with lanthanum (57), lutetium (71), actinium (89), and lawrencium (103). Some sources put La and Ac where Lu and Lr are on the tear card, some sources have it the way it's shown on the tear card, and some even put Lu and Lr at the bottom of the periodic table and leave La and Ac where they are! What's it all mean? It means that chemists, like the rest of us, can't always agree on how to do things. Don't worry—as long as you know the basic ideas that we've been talking about, you're in good shape!

The Four Main Periodic Trends

One of the things that keep coming up during this chapter is the octet rule. Since it's really, really important, we'll go over it again.

Figure 7.1

The octet rule is the driving force for chemical reactions and properties. If you learn only one thing from this chapter, learn the octet rule!

OCTET RULE: ALL ELEMENTS GAIN OR LOSE ELECTRONS SO THEY WIND UP WITH THE SAME ELECTRON CONFIGURATION AS THE NEAREST NOBLE GAS.

All elements follow the octet rule because having completely filled s- and p-orbitals in the outermost energy level makes elements very stable. As we saw with the alkali metals and halogens, elements that are only one electron away from a noble gas electron configuration are extremely reactive.

As a demonstration of how this works, imagine taking a long car trip with a very small, very hungry child. Let's say it's six o'clock and you see a sign that says, "Stop at Burger World, only 95 miles ahead!" and you know the child will demand to get a burger. When you tell the kid that Burger World is about an hour and a half ahead, the kid will whine a moment and settle back down to pulling his sister's hair. Although he's hungry, he knows he'll have to wait. This is similar to how nitrogen feels about electrons—though it wants to gain three electrons to have the same electron configuration as neon, it can wait.

Bad Reactions

Chemistry teachers frequently get annoyed when you say, "nitrogen wants to …" in reference to a chemical reaction. Many will rightly tell you that atoms don't have any particular desires. During the course of this book, when I say that an atom "wants to do [something]", what I really mean is "the atom will become more stable when it does [something]."

Now, let's imagine that you drive a while and the sign now says "Burger World, next exit!" If you're unfortunate enough to have a halfway-literate child, the screaming from the back seat will rise to a fevered pitch until hamburgers come flying into the child's mouth. This is roughly how fluorine feels about electrons—since it's so close to having a noble gas configuration, it reacts violently so it can get that extra electron *now!*

In this section, we'll talk about some properties that can be predicted based on an element's position in the periodic table. When going over the following trends, keep in mind that the sole reason that atoms ever do anything is to become more stable.

Ionization Energy

The ionization energy of an element is the amount of energy required to pull one electron off of an atom. The units for ionization energy are kilojoules per mole, or kJ/mol. (We'll discuss what a "mole" is in Chapter 11.) The ionization process is described by the following figure:

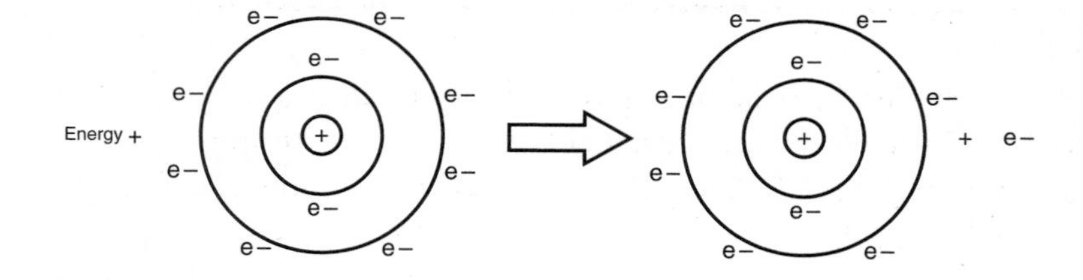

Figure 7.2

When enough energy is added to an atom, an electron can be removed. This process is called ionization.

As you move from left to right across the periodic table, the energy required to remove an electron increases because of the octet rule. When you think about how the octet rule works, this should make sense. Remember that all elements want to gain or lose electrons to be like the nearest noble gas. Elements on the left side of the periodic table (such as lithium) want to lose electrons to be like the nearest noble gas. As a result, it doesn't take much energy to pull an electron off of the atom, because it wants to lose the electron anyway.

Elements on the right side of the periodic table (such as fluorine) want to gain electrons, not lose them, so it takes a very large amount of energy to pull off an electron. The highest ionization energies belong to the noble gases, which are the most stable due to their filled s- and p-orbitals and therefore don't want to lose any electrons.

Ionization energy decreases as you move down a family of elements. The reason for this is that electrons in low-energy levels repel electrons in high-energy levels away from the nucleus because both have negative charge. This phenomenon is called the *shielding effect*, and is shown in the following figure:

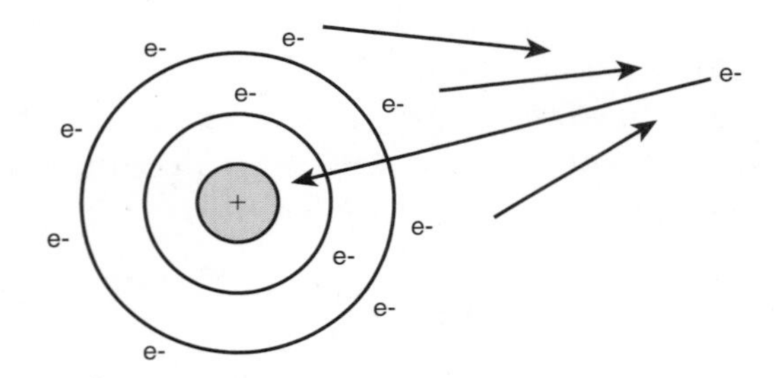

Figure 7.3

Inner electrons tend to push outer electrons away from the nucleus because both have negative charge.

Though outer electrons are attracted to the nucleus because they have opposite charges, they are simultaneously pushed away from the nucleus due to repulsions with electrons from inner energy levels. As a result of this shielding effect, outer electrons are less tightly bound to the nucleus than inner electrons.

Chemistrivia

So far, we've discussed the ionization energy associated with pulling off only one electron from an atom. This value is called, simply enough, the first ionization energy. If we were to pull off additional electrons, the energy required to pull off the second electron would be called the second ionization energy, the third would be the third ionization energy, and so on.

Electron Affinity

Electron affinity is the energy change that occurs when a gaseous atom picks up an extra electron. The units for electron affinity are kilojoules per mole (kJ/mol). This process is shown by the following figure.

Figure 7.4

Because energy is given off when an atom picks up an electron, electron affinities are always negative values.

Lithium, as might be expected, doesn't want to pick up an extra electron. The reason for this is that the octet rule causes lithium to want to lose electrons, not gain them. As a result, not much energy is released when lithium is forced to pick up an electron, causing the electron affinity to be only slightly negative. Fluorine, on the other hand, wants very much to pick up an extra electron to be like neon. Because a lot of energy is given off when fluorine does this, its electron affinity is very negative. Generally, electron affinity becomes more negative as you move left to right across the periodic table.

As you move down a family in the periodic table, elements, in general, want to gain electrons less because of the shielding effect. As a result, elements at the bottom of the periodic table tend to have less negative electron affinities than those at the top.

The Mole Says

The noble gases don't have electron affinities because they don't want to pick up any extra electrons. After all, they're already stable!

Electronegativity

Electronegativity is very similar to electron affinity, in that both measure how easy it is for an atom to gain electrons. Whereas electron affinity deals with isolated atoms in the gas phase, electronegativity is a measure of how much an atom will tend to pull electrons away from other atoms that it has bonded to.

As you might imagine, the trend for electronegativity follows the trend for electron affinity fairly closely. Because the octet rule states that elements on the left side of the periodic table want to lose electrons while those on the right want to gain electrons, electronegativity increases as you move from left to right across the periodic table. Remember, however, that the noble gases have no electronegativity because they tend not to gain electrons at all. Likewise, the shielding effect causes the electronegativity of elements to decrease as you move down a family in the periodic table.

Chemistrivia

Unlike electron affinity and ionization energy, there is no unit associated with electronegativity. Electronegativities are assigned via a method developed by Linus Pauling in 1932 and are based on the relative bond energies of different elements. For this and other research about chemical bonds, Pauling won the 1954 Nobel Prize in chemistry (his first of two Nobel Prizes).

Atomic Radius

Atomic radius seems like it should be an easy thing to define. After all, it's just the radius of an atom, so it measures how big an atom is. No problem, right?

Wrong! Remember that quantum mechanics doesn't actually set a size limit on the atom, so atoms are, theoretically speaking, infinitely large. As a result, we can't just get a very tiny ruler and find the radius of one.

So, how do we do find the radius of an atom? One of the most common ways is by examining two atoms of the same element that have bonded to one another. The midpoint between the two atoms is where one of the atoms stops and the other starts. Our definition for atomic radius is one half the distance between the nuclei of two bonded atoms of the same element.

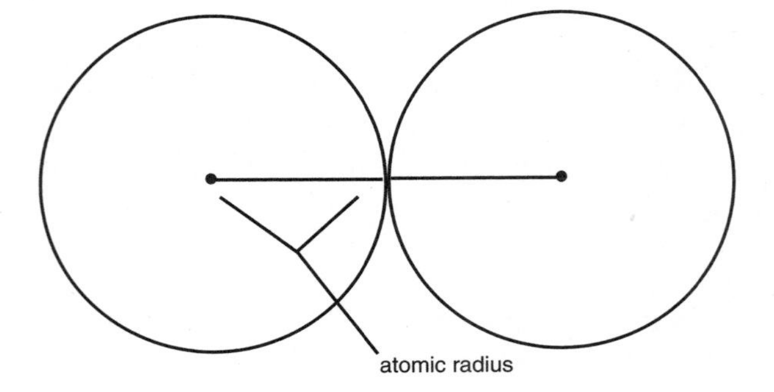

Figure 7.5

Because both of the atoms are the same, we can arbitrarily say that the halfway point between them is where one atom stops and the other starts.

As you move from left to right across the periodic table, the atomic radius of the elements decreases. This may seem counterintuitive, because every element has one more electron than the one before it. Shouldn't a larger number of electrons cause the atom to be larger?

Not necessarily. Remember that the atoms within a period all have pretty much the same energies. Though additional electrons are added, they don't have additional energy. However, at the same time we're adding electrons, we're also adding protons (every element also has one more proton than the one before it). Though the electrons have the same energies, the additional protons cause the nucleus to pull more tightly on the electrons. As a result of this increased nuclear attraction for the electrons, the atomic radii of atoms on the right side of the periodic table are less than those of atoms on the left side.

As you move down a family on the periodic table, the atomic radii of atoms increase. This increase is relatively easy to understand, because every row has one more energy level than the one before it. Because the electrons are in higher energy levels, they are better able to pull away from the nucleus, making the atomic radii larger.

You've Got Problems

Problem 1: Arrange these elements from lowest to highest electronegativity and atomic radius: O, F, P, Rb, Sn.

The Least You Need to Know

- It is possible to predict the properties of an element based on its location in the periodic table.
- The octet rule states that all elements tend to gain or lose electrons in order to have the same electron configuration as the closest noble gas. Noble gases are particularly stable because their outer s- and p-orbitals are completely filled.
- Periodic trends such as electronegativity, ionization energy, and electron affinity are based on both the shielding effect and the octet rule.

Ionic Compounds

In This Chapter

- The properties of ionic compounds
- How ionic compounds are formed
- Properties of ionic compounds
- Naming ionic compounds

It may have crossed your mind that you've gotten seven chapters into a book about chemistry without actually ever talking about chemical compounds. Sure, I defined what a chemical compound was in Chapter 6, but I never really explained it further than that.

As you might have guessed, chemical compounds are far more numerous and common than pure elements. The coffee you drink is a mixture of many different chemical compounds, as is the hamburger you had for lunch and the peanut brittle you had for a snack last night while watching monster movies at 3 A.M. Okay, that last one was what *I* ate yesterday morning at 3 A.M., but you get the idea.

Unlike elements, there's no periodic table of the compounds to make them easier to work with. However, with a little more knowledge, you'll find that the regular periodic table of elements gives you just about everything you need to know about chemical compounds. In this chapter, we'll talk about one of the most common types of compounds—ionic compounds.

What's an Ionic Compound?

In Chapter 7, we talked about the octet rule, which states that "All elements gain or lose electrons so they wind up with the same electron configuration as the nearest noble gas." Basically, what this means is that a neutral atom of any element other than a noble gas isn't very stable. As a result, it will gain or lose electrons until it attains the stable electron configuration of a noble gas. Atoms that have gained electrons are called *anions* and have negative charge, and atoms that have lost electrons are called *cations* and have positive charge. When anions stick to cations, the resulting chemical compound is called an *ionic compound*.

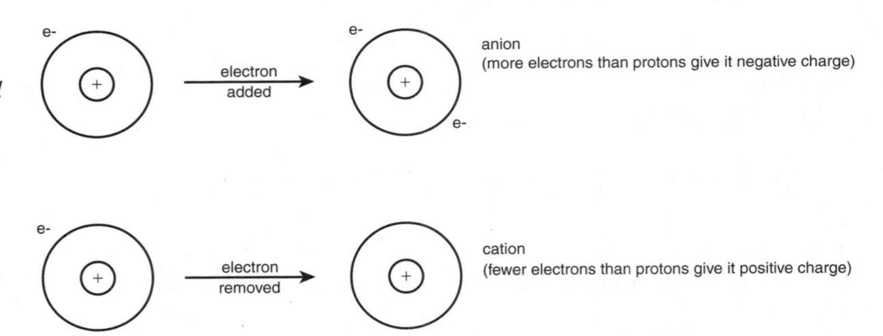

Figure 8.1

Anions and cations are formed when an atom gains or loses electrons to achieve the electron configuration of the closest noble gas.

One question you may be asking yourself is, "How many electrons do different elements want to gain or lose?" The answer to this question, like many things in chemistry, can be found on the periodic table.

The best way to figure out how many electrons will be gained or lost by a specific element is to count forward from it in the periodic table until you reach the next noble gas, and then count backward from it in the periodic table until you reach the last noble gas. If the forward direction requires less counting than the backward direction, the element will gain the same number of electrons as you counted to the closest noble gas to form an anion. Likewise, if the backward direction requires less counting, the element will lose the same number of electrons as you counted to the nearest noble gas to form a cation. Now, this is very important: *Skip over the transition metals when counting to the noble gases*—otherwise, things won't work out the way you'd like.

def•i•ni•tion

An **anion** is a negatively charged atom or group of atoms, while a **cation** is an atom or group of atoms with positive charge. When an anion sticks to a cation, the result is an **ionic compound**.

For example, oxygen forms ionic compounds as an anion with –2 charge because it needs to gain two electrons to have the same electron configuration as neon. Gallium, on the other hand, has a +3 charge because it needs to lose three electrons to have the same electron configuration as argon. If you're confused because it looks like you need to count backward by 13 rather than three, remember the rule to ignore the transition metals!

You've Got Problems

Problem 1: What will the charges of the following elements be when they gain or lose electrons to gain the same electron configurations as the nearest noble gas?

(a) magnesium (Mg)

(b) aluminum (Al)

(c) bromine (Br)

How Ionic Compounds Are Formed

As you might expect from the previous section, the octet rule plays a huge role in ionic compound formation. Let's imagine what happens when lithium reacts with chlorine to form an ionic compound.

Lithium has very low electronegativity, meaning that it tends not to want electrons. Chlorine, on the other hand, has a very high electronegativity, meaning that it wants to gain electrons. As a result, lithium gives electrons to chlorine when one atom of each comes into contact with the other.

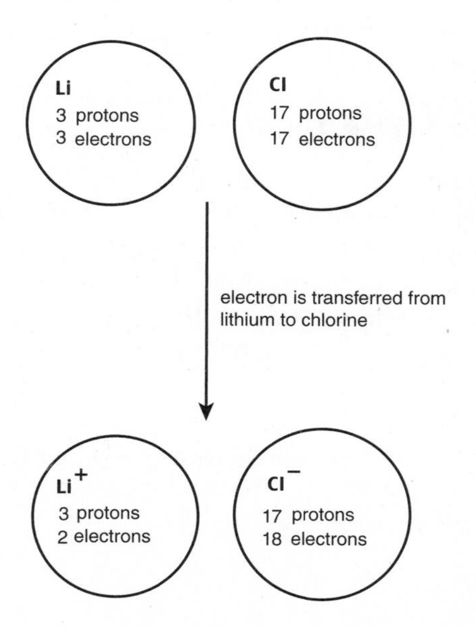

Figure 8.2

Chlorine's high electronegativity causes it to pull electrons from lithium, resulting in the formation of the ionic compound LiCl.

When this occurs, lithium goes from having no charge to a +1 charge, while chlorine goes from neutral to having a –1 charge. Because the lithium cation and chlorine anion have opposite charges, they attract one another and form lithium chloride, LiCl.

Generally, ionic compounds are formed whenever two elements with very dissimilar electronegativities (greater than 2.1) bond with each other. As a result, oxygen (electronegativity = 3.4) will form an ionic compound with lithium (electronegativity = 1.0) because the difference in electronegativity between these elements is 3.4 – 1.0 = 2.4. Oxygen, however, does not form ionic compounds with nitrogen (electronegativity = 3.0) because their electronegativities are so similar. Because metals and nonmetals frequently have such dissimilar electronegativities, it's usually a good guess that compounds formed by the combination of metals and nonmetals are ionic.

You've Got Problems

Problem 2: Based on their positions in the periodic table, determine which of the following compounds are ionic:

(a) BaF_2

(b) SiO_2

(c) N_2

Properties of Ionic Compounds

Because all ionic compounds are formed when anions and cations are attracted to one another, ionic compounds frequently have similar characteristics.

Ionic Compounds Form Crystals

Ionic compounds consist of cations and anions that stick next to each other because of their opposite charges. Imagine a single lithium cation stuck next to a single chlorine anion to form lithium chloride. Now, it's unlikely that only one lithium ion and one chloride ion will be present in this location—generally, when we speak of chemical reactions, we're talking about a huge number of atoms undergoing a reaction in a very small place (one teaspoon of salt contains approximately 10^{22} atoms). As a result, if our single LiCl pair were to come close to another LiCl pair, the following would take place:

Because oppositely charged ions attract one another, the LiCl pairs will tend to form larger groups. These larger groups, in turn, will form even larger groups of ions, as shown in the following figure.

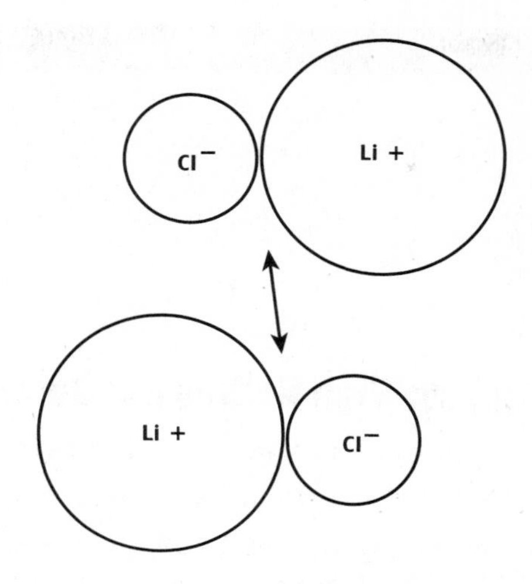

Figure 8.3

The positive charge on the lithium cation of one pair will be attracted to the negative charge on the chloride ion of the other pair.

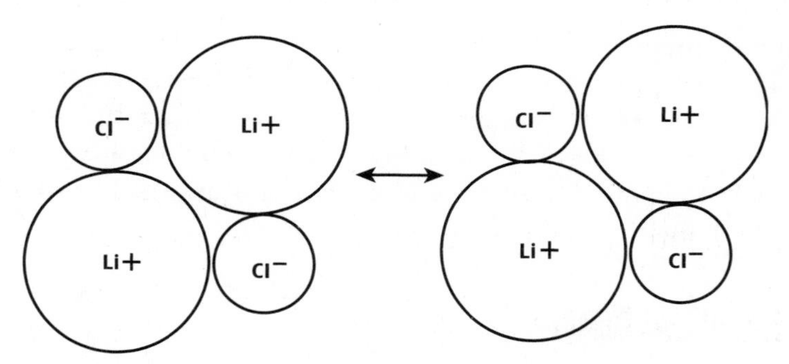

Figure 8.4

This process, where stacks of LiCl ions combine with one another, will continue until there are no more lithium or chloride ions.

These large arrangements of ions are referred to as *crystals*. Though crystals are frequently formed from ionic compounds, they also exist in some other chemical compounds, such as diamonds. We'll talk about this in much greater detail in Chapter 12.

def•i•ni•tion

Crystals are large arrangements of ions or atoms that are stacked in regular patterns. Many ionic compounds form very large crystals.

The Mole Says

The word "ionic compound" is synonymous with "salt." To distinguish sodium chloride from other salts, chemists often refer to it as "table salt" rather than by the generic term "salt."

Chemistrivia

Hydrates are formed when one or more molecules of water attach themselves to ionic compounds. These compounds are interesting because they appear dry but give off water when heated. Particularly interesting is Epsom salt, or magnesium sulfate heptahydrate ($MgSO_4 \cdot 7H_2O$). When heated, enough water is given off that it actually dissolves the magnesium sulfate!

Ionic Compounds Often Have High Melting and Boiling Points

What happens when you heat something up in your kitchen? You may have discovered while cooking (or while microwaving random things while bored) that most of the foods we eat either melt or burn when heated. Some foods even do both! As you can probably guess, I'm an expert when it comes to putting out house fires.

Ionic compounds, on the other hand, frequently melt and boil at much higher temperatures than other materials. In order for ionic compounds to melt, enough energy must be added to make the cations and anions move away from one another. Because these attractions are so strong, it takes a lot of energy to pull these ions apart. Adding this much energy to ionic compounds requires a great deal of heat, which is why ionic compounds have very high melting and boiling points.

Ionic Compounds Are Hard and Brittle

Imagine bashing a big chunk of lithium chloride against your head. What do you suppose that might feel like? If you guessed that it would hurt like crazy, you were right. Like many ionic compounds, lithium chloride is as hard as a rock.

Ionic compounds are extremely hard because it is difficult to make the ions move apart from each other in a crystal. Even if you apply a great deal of force on the crystal (imagine running headlong into a giant wall of lithium chloride), the attraction between the cations and anions will frequently continue to hold the crystal together.

Let's say, though, that you really want to break apart an ionic compound. While very hard, ionic compounds are also frequently very brittle, meaning that they break apart when the right kind of force is applied. As the following figure shows, where you apply the force is just as important as how much force you use.

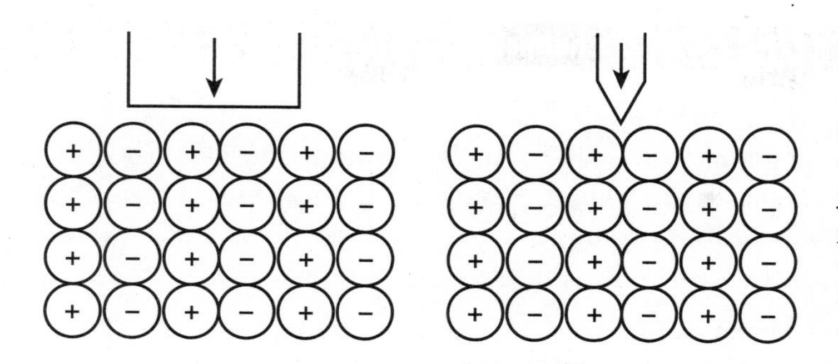

Figure 8.5

By applying force in a way that pries the cations and anions apart from each other: you can cause a crystal to completely break apart. In this case, the crystal on the left will hold together and the crystal on the right will likely shatter.

As you can see from this diagram, ionic crystals align themselves such that there are regions where a small force can break apart the crystal. These regions are sometimes referred to as "cleavage planes" because they are the locations where the crystal is weakest and can most easily be broken.

Ionic Compounds Conduct Electricity When Dissolved in Water or Melted

Once upon a time, there was an inventor who came up with a device for drying hair. This "hairdryer" as he called it, heated air with electricity and blew it across the hair of the person holding it. Because water evaporates when heated, the hair dried more quickly. This inventor's legacy lives on to this day in a household appliance loved by millions.

Shortly afterward, there was a guy who decided that he didn't want to wait to get out of the bathtub before drying his hair. His legacy: A hairdryer warning sticker with a picture of a guy getting electrocuted.

When ionic compounds are placed in water, they cause the water to conduct electricity. Normally, water doesn't conduct electricity well at all. However, when salts dissolve in water, they break up into their constituent cations and anions and it is the presence of these ions that allows it to conduct electricity. Because salts conduct electricity when dissolved in water, they are referred to as *electrolytes*.

In the same way, pure salts also conduct electricity when they are melted. As a solid, the anions and cations in an ionic compound are locked in place and unable to move electrical charge. However, when the ionic compound is melted, these ions are free to move around and conduct charge.

def•i•ni•tion

Electrolytes are compounds that conduct electricity when dissolved in water. Many ionic compounds are considered to be electrolytes. However, some ionic compounds don't dissolve in water. As a result, they don't share this property.

What's in a Name? Ionic Nomenclature

If you've been doing chemistry for any length of time, you're well aware of the difficulty posed with naming ionic compounds. Anybody who has needed $CuCl_2$ and found that the only chemicals available are "cuprous chloride" and "cupric chloride" knows how much trouble naming ionic compounds can be. Fortunately, we're going to discuss a method for making ionic naming simple and easy.

Before we get started with naming, we need to learn about polyatomic ions. As the name suggests, polyatomic ions are ions that contain more than one atom. An example of a polyatomic ion is the hydroxide ion, which has the formula OH^{-1}; together, the oxygen and hydrogen have a net –1 charge. Generally, polyatomic ions are anions, though the ammonium and mercury(I) cations are exceptions. Polyatomic ions are found in many ionic compounds, making it important that you learn the names and formulas of the more common ions.

Name of Ion	Formula and Charge of Ion
acetate	$C_2H_3O_2^{-1}$
ammonium	NH_4^{+1}
bicarbonate	HCO_3^{-1}
bisulfate	HSO_4^{-1}
carbonate	CO_3^{-2}
chlorate	ClO_3^{-1}
chlorite	ClO_2^{-1}
chromate	CrO_4^{-2}
cyanide	CN^{-1}
dichromate	$Cr_2O_7^{-1}$
hydroxide	OH^{-1}
mercury(I)	Hg_2^{2+}
nitrate	NO_3^{-1}
nitrite	NO_2^{-1}
permanganate	MnO_4^{-1}
phosphate	PO_4^{-3}
sulfate	SO_4^{-2}
sulfite	SO_3^{-2}

Writing Ionic Names from Formulas

Now that we're familiar with polyatomic ions, let's learn how to name ionic compounds when given their chemical formulas by using the following steps.

Step 1

Determine the "base name" of the ionic compound. Ionic compound base names contain two words:

- The first word is the name of the cation. Unless the cation is "ammonium" (in which case you already know its name), the name of the cation is the same as the name of the element. For example, the first word in "NaOH" is "sodium."
- The second word is the name of the anion. If the anion is a polyatomic ion, you can just look up the name on the chart of polyatomic ions shown earlier. For example, "NaOH" is "sodium hydroxide." If the anion is a single element, replace the ending of the element name with "-ide." For example, NaBr would be "sodium bromide."

Step 2

Determine whether or not the compound will require a Roman numeral.

For many compounds, you can stop with the base name. However, some elements that form cations can have more than one possible charge. For example, iron can form two ionic compounds with chlorine; "$FeCl_2$" and "$FeCl_3$." Because the naming system we just learned would call both of these compounds "iron chloride," we need some way to distinguish between them.
To do this, we write a Roman numeral after the name of the cation to indicate the amount of positive charge it has.

Unfortunately, we can't just go putting Roman numerals for all ionic compounds. We can only do it for compounds containing cations that commonly bear more than one possible charge. To help us with this job, take a look at the following figure, which shows the positive charges of the most common transition metal cations.

Bad Reactions

Use Roman numerals only when naming ionic compounds that have cations with more than one possible positive charge. If you place Roman numerals in all compound names, they will be wrong when misapplied.

Figure 8.6

Use this diagram to determine which elements have only a single possible charge and which elements can exist with more than one possible positive charge.

										13 Al +3			
21 Sc +3	22 Ti +2, +3, +4	23 V +2, +3, +4, +5	24 Cr +2, +3, +6	25 Mn +2, +3, +4, +6 +7	26 Fe +2, +3	27 Co +2, +3	28 Ni +2, +3	29 Cu +1, +2	30 Zn +2	31 Ga +3	32 Ge +2, +4		
39 Y +3	40 Zr +4	41 Nb +3, +5	42 Mo +6	43 Tc +4, +6, +7	44 Ru +3	45 Rh +3	46 Pd +2, +4	47 Ag +1	48 Cd +2	49 In +3	50 Sn +2, +4	51 Sb +3, +5	
71 Lu +3	72 Hf +4	73 Ta +5	74 W +6	75 Re +4, +6, +7	76 Os +3, +4	77 Ir +3, +4	78 Pt +2, +4	79 Au +1, +3	80 Hg +1, +2	81 Tl +1, +3	82 Pb +2, +4	83 Bi +3, +5	84 Po +2, +4
103 Lr +3	104 Rf	105 Db	106 Sg	107 Bh	108 Hs	109 Mt							

57 La +3	58 Ce +3, +4	59 Pr +3	60 Nd +3	61 Pm +3	62 Sm +2, +3	63 Eu +2, +3	64 Gd +3	65 Tb +3	66 Dy +3	67 Ho +3	68 Er +3	69 Tm +3	70 Yb +2, +3
89 Ac +3	90 Th +4	91 Pa +4, +5	92 U +3, +4, +5, +6	93 Np +3, +4, +5, +6	94 Pu +3, +4, +5, +6	95 Am +3, +4, +5, +6	96 Cf +3	97 Bk +3, +4	98 Cf +3	99 Es +3	100 Fm +3	101 Md +2, +3	102 No +2, +3

As you can see, some elements can have several different possible positive charges. Cobalt (Co), for example, can have either a charge of +2 or +3, making it necessary to use Roman numerals to distinguish between them. Zinc, on the other hand, doesn't require Roman numerals in its compound names because it only forms stable cations with a charge of +2.

Step 3

Determine the Roman numeral that goes after the cation name.

To do this, use the following formula:

$$\text{Roman numeral} = \frac{-(\text{charge on anion}) \times (\text{number of anions})}{(\text{number of cations})}$$

Let's see how that works with the examples $FeCl_2$ and $FeCl_3$, both of which have the base name "iron chloride."

$FeCl_2$ contains two chloride ions, each of which has a negative charge of 1. Because there is no subscript under the Fe, there is only one iron atom present. As a result, the Roman numeral required for $FeCl_2$ is: $\frac{-(-1)(2)}{1} = 2$, making this compound iron(II) chloride.

$FeCl_3$ contains three chloride ions with negative charges of 1. Because there is only one iron atom, the Roman numeral for $FeCl_3$ is: $\frac{-(-1)(3)}{1} = 3$

The name of $FeCl_3$ is iron(III) chloride.

You've Got Problems

Problem 3: Name the following ionic compounds using the previous rules:

(a) Na_2SO_4

(b) Cu_2O

(c) $CoCO_3$

(d) NH_4Cl

Writing Ionic Formulas from Names

As you might have guessed, writing formulas from names is pretty much the reverse process of writing names from formulas. Let's learn the steps.

Step 1

From the base name, determine the formula and charge of the ions.

Let's say we were told to write the formula for "calcium sulfate." From the name "calcium" we know that the cation will be Ca^{+2}. The "Ca" part is simply the atomic symbol for calcium, and the "+2" is given to us by the octet rule because calcium needs to lose two electrons to get the same electron configuration as argon.

Because sulfate isn't an element on the periodic table, many people start screaming in panic and confusion. If you get into a situation where an unfamiliar ion shows up, take a look at the chart of polyatomic ions and see if the ion is listed there. As it turns out, sulfate is the SO_4^{-2} ion. Feel better?

Step 2

Write the formulas of the cations and the anions next to each other.

This isn't so bad! In our example, we just write Ca^{+2} SO_4^{-2}.

Step 3

Devise an ionic formula that gives the compound a neutral charge.

In our example, the charges on the calcium cation and sulfate anion cancel each other. As a result, the compound will be electrically neutral when one calcium ion combines with one sulfate ion, forming $CaSO_4$.

The Mole Says

If you have more than one polyatomic ion, you must always put parentheses around it before placing the subscript under it. "Beryllium hydroxide" is $Be(OH)_2$, not $BeOH_2$. However, if the ion is not polyatomic, never use parentheses!

Let's go through another example: beryllium hydroxide.

1. "Beryllium" indicates Be^{+2} and "hydroxide" indicates OH^{-1}.
2. Putting them together, we get Be^{+2} OH^{-1}.
3. Because beryllium hydroxide has to be electrically neutral, there need to be two hydroxide ions for each beryllium ion. As a result, the formula of beryllium hydroxide is $Be(OH)_2$.

You've Got Problems

Problem 4: Write the formulas of the following ionic compounds:

(a) lithium acetate

(b) sodium nitrate

(c) chromium(VI) sulfate

(d) zinc phosphate

The Least You Need to Know

- Ions are formed when atoms gain or lose electrons because of the octet rule.
- Atoms with low electronegativities give electrons to atoms with high electronegativities so both can gain the electron configuration of the nearest noble gas. This causes them to form ions with opposite charges, which are attracted to one another to form ionic compounds.
- The properties of ionic compounds are primarily determined by the strong attractions between many anions and cations.
- Naming ionic compounds isn't really all that hard if you use a methodical system.

Getting to Know Covalent Compounds

In This Chapter

- What are covalent compounds?
- How are covalent compounds formed?
- Properties of covalent compounds
- Naming covalent compounds

Now that we know everything (well, almost everything) there is to know about ionic compounds, we're ready to broaden our horizons to other types of chemicals. It's time to spread our metaphorical wings and fly to the land of the covalent compound!

Okay, maybe that last paragraph was a little overdone. However, the idea is pretty much true. In this chapter, we're going to learn almost everything about covalent compounds—what they are, what their properties are, and how to name them.

What Are Covalent Compounds?

In Chapter 8, we learned that ionic compounds are formed when an electronegative atom grabs an electron from an atom with low electronegativity. The reason for this is the octet rule, which states that all elements want to gain or lose electrons so they have the same electron configuration as the closest noble gas.

One thing we didn't discuss, however, is what happens when two electronegative atoms react with one another. For example, both nitrogen and hydrogen want to gain electrons to be like their nearest noble gas—this suggests that they won't give electrons to one another. By doing so, one would actually be further away from its goal of a full electron shell. Surprisingly, hydrogen and nitrogen actually form a large number of chemical compounds with one another, including everybody's favorite household cleaner, ammonia (NH_3). How does this work, anyway?

I'm glad you asked! Let's imagine the following scenario:

Iodine wants another electron to be like its nearest noble gas, xenon. Hydrogen also wants another electron to be like its nearest noble gas, helium. What happens when a neutral hydrogen atom bumps into a neutral iodine atom? They do exactly what we were all taught to do in kindergarten—they share!

Let's take a quick break from this example to define a couple of terms that will enable us to understand how this is relevant in the real world of atoms and electrons:

- *Valence electrons* are the number of s- and p-electrons past the most recent noble gas. For example, lithium has one valence electron and nitrogen has five valence electrons. These valence electrons are important because all elements tend to want to have the same number of valence electrons as the nearest noble gas (our buddy the octet rule), and these valence electrons determine the reactivities of many chemical compounds. Elements generally want a total of eight valence electrons to fill their outermost s- and p-orbitals; important exceptions are hydrogen, beryllium, and lithium, each of which wants two valence electrons to be like helium.
- *Covalent bonds* are the bonds formed when two atoms share a pair of valence electrons. All covalent bonds contain two electrons.

def•i•ni•tion

Valence electrons are the number of s- and p-electrons since the most recent noble gas. These electrons are primarily responsible for the reactivities of the elements in the s- and p-sections of the periodic table. **Covalent bonds** are the bonds created when two valence electrons are shared.

Back to our example:

Iodine has seven valence electrons, making it one short of the eight valence electrons it needs in order to be like xenon. Hydrogen has one valence electron, making it one short of the two valence electrons it needs in order to be like helium.

Here's the neat part: If hydrogen and iodine share their valence electrons, they can both pretend that they have the same number of valence electrons needed to be like their nearest noble gases. Here's a drawing of what this looks like.

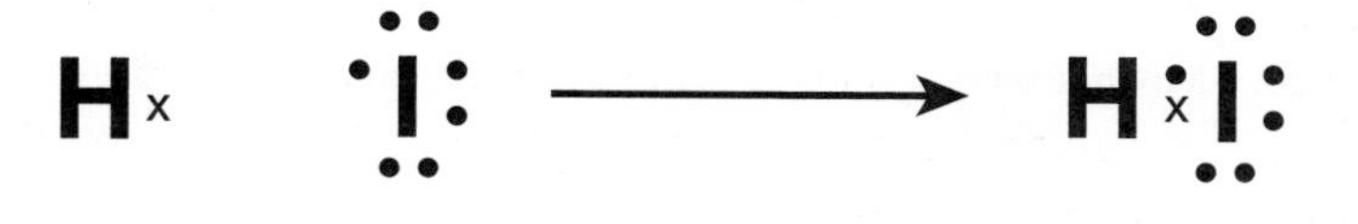

Figure 9.1

Before iodine and hydrogen combine, they're both missing the one valence electron they need to have the same electron configuration as the nearest noble gas. By sharing their unpaired valence electron, each pretends it has the right number of valence electrons—by doing so, they form a covalent bond.

From the preceding figure, we can see that both atoms simulate having the right number of valence electrons by sharing their unpaired electron with the other. The left side of the figure shows that iodine has only seven valence electrons. However, once it bonds with hydrogen, it has eight valence electrons around it. Of course, the number of electrons doesn't change, but any shared electrons count toward both the valence electrons for both atoms. Likewise, hydrogen has two valence electrons around it, making it stable.

Let's see what happens when hydrogen combines with oxygen to form water, H_2O.

Each hydrogen atom has only one valence electron. In order to get the desired two valence electrons, each needs to gain another electron. Oxygen, however, has only six valence electrons. To be like neon with its eight valence electrons, it needs to gain an additional two. The following figure illustrates how one oxygen atom and two hydrogen atoms bond to form H_2O.

Chemistrivia

Oxygen contains two pairs of electrons that don't bond at all. These electron pairs are referred to as "unshared electron pairs," "lone pairs," or "unbonded pairs."

Figure 9.2

When two hydrogen atoms share one electron each with an oxygen atom, the three atoms form a chemical compound with two covalent bonds.

The Mole Says

You may have noticed that this figure initially shows the electrons on oxygen spread out on all four sides rather than fully paired in three positions. We show electrons unpaired whenever possible because Hund's rule states that electrons prefer to remain unpaired (Chapter 5).

You've Got Problems

Problem 1: Sketch what will happen when an atom of nitrogen combines with three atoms of hydrogen to form ammonia (NH_3).

As you can see, oxygen has two unpaired electrons that need to be paired up in order to be like neon. As in our previous example, each hydrogen atom needs one more electron to be like helium. The problem is solved when both hydrogen atoms form covalent bonds with oxygen, forming H_2O.

Formation of Multiple Covalent Bonds

The examples we just discussed explain why single covalent bonds are formed between two atoms. However, two atoms sometimes bond more than once. How does this work?

Let's use the example of O_2. Both oxygen atoms have six valence electrons, meaning that they each need two more to be like neon. Fortunately, by combining both sets of unpaired electrons simultaneously, they both achieve their desired electron configurations.

Figure 9.3

By combining more than one unpaired electron at a time, a double bond is formed, and both oxygen atoms end up with eight valence electrons.

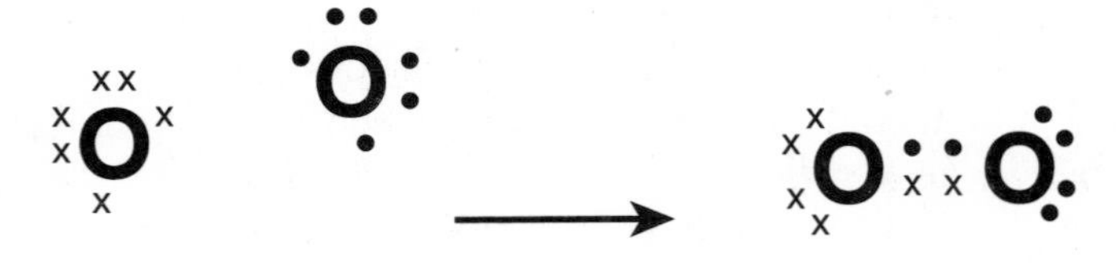

Because the two oxygen atoms are sharing four electrons at the same time, two covalent bonds are formed between them. We refer to these two covalent bonds as being a "double bond." In the same way that this double bond was formed, you can imagine triple bonds being formed if two atoms each have three unpaired electrons. An example is nitrogen (N_2).

Bad Reactions

It's common to assume that if an atom forms a triple bond in one compound, it will always form triple bonds in covalent compounds. Remember that each example is unique and requires separate analysis.

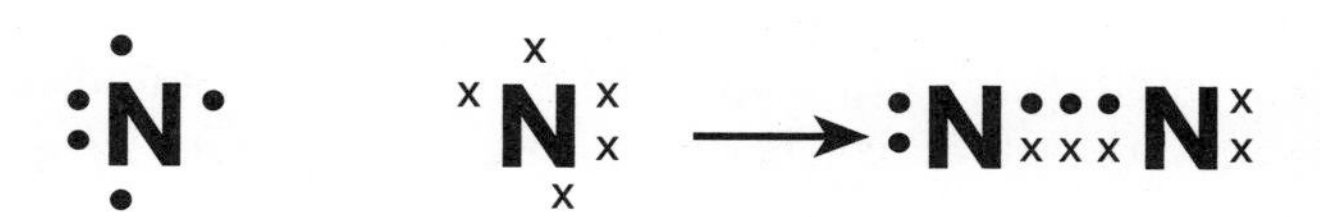

Figure 9.4

When two atoms that each have three unpaired electrons combine with each other, the result is a triple bond.

Properties of Covalent Compounds

When we talked about solid ionic compounds in Chapter 8, we found that their properties often derive from the strong attraction of opposite electrical charges. It should not come as a surprise to find that the properties of covalent compounds are largely owing to the nature of covalent bonds.

One of the most important things to remember about covalent compounds is that they're not ionic. This seems obvious, but the difference is actually subtler than you may imagine. To illustrate this concept, take a look at the following figure.

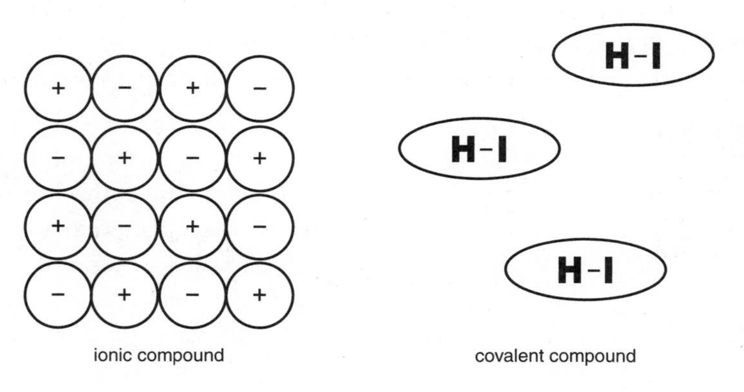

Figure 9.5

The properties of solid ionic compounds are based on the fact that many ions are rigidly held in place with electrical forces. Molecules in covalent compounds, however, operate with relative independence from neighboring molecules.

Unlike ionic compounds, where all of the ions in a large crystal help to hold each other together, the molecules in a covalent compound are held together by forces called "intermolecular forces," which are much weaker than chemical bonds (more about intermolecular forces in Chapter 12). As a result, the molecules in a covalent compound are not attracted to each other as much as the ions in ionic compounds. This difference in structure is important in understanding the properties of covalent compounds.

Covalent Compounds Have Low Melting and Boiling Points

As mentioned in Chapter 8, a large amount of energy is required to melt an ionic compound because of the strong interactions between the cations and anions in an ionic crystal. However, in covalent compounds, all molecules are bound only weakly to neighboring molecules; therefore, it takes very little energy to separate covalent molecules from one another.

Bad Reactions

Many beginning chemistry students falsely believe that when a covalent molecule melts, covalent bonds are broken. This is false. When ionic compounds melt, the ionic attraction fails. When covalent compounds melt, the molecules simply pull away from each other, leaving the bonds intact.

Covalent Compounds Are Poor Conductors

Ionic compounds are great conductors of electricity when dissolved or melted. As mentioned in Chapter 8, this is because ionic compounds have mobile ions that are able to transfer electrical charge from one place to another. They also conduct heat very well because the ions are all right next to each other, making it possible for energy to be transferred efficiently from one place to another.

Chemistrivia

Since water is a covalent compound, you may be wondering why it conducts electricity so well. The fact of the matter is that it doesn't conduct electricity well at all. It's only when ionic compounds are dissolved in it (as is almost always the case) that it's a good conductor.

Covalent compounds, on the other hand, are almost always good insulators of both electricity and heat. Electricity is not able to conduct efficiently through covalent compounds because there are no ions to move the electrical charge. An excellent example of this is in your own house, where the metal in your extension cords is covered with plastic to avoid electrocuting your cat. Heat also doesn't travel well through covalent compounds because the molecules aren't as tightly held to each other as the ions in an ionic compound, making heat transfer less efficient.

This is why you use oven mitts to take your cookies out of the oven rather than coating your hands with salt.

Covalent Compounds Sometimes Burn

Many covalent compounds are flammable and burn readily with the addition of heat. The main group of covalent compounds that are flammable are called *organic compounds*. Organic compounds are covalently bonded carbon atoms that burn because they contain carbon and hydrogen, both of which combine nicely with oxygen at high temperatures.

> **def•i•ni•tion**
>
> **Organic compounds** are covalent compounds that contain carbon and hydrogen. They may also contain smaller amounts of other elements such as nitrogen, sulfur, phosphorus, oxygen, or any of the halogens.

It's important to keep in mind that not all covalent compounds burn—for example, water is a covalent compound and you'll have a very hard time starting a fire with it. However, many more covalent than ionic compounds are flammable.

Flammability is a general property of covalent compounds because a large majority of the known covalent compounds are organic. Since most organic compounds burn, we can safely list this as a property of covalent compounds even though there are many covalent compounds that don't burn.

What's in a Name? Covalent Nomenclature

Fortunately, it's much easier to name many covalent compounds than it is to name ionic compounds (see Chapter 8). Organic compounds have a separate naming system that we'll discuss in Chapter 25, so if you've seen things like benzene and 3-methylhexane, don't worry about them just yet.

Naming Covalent Compounds from Formulas

Covalent compounds have two names. The following rules will enable us to name nonorganic covalent compounds with the greatest of ease.

Rule 1

The first word is the name of the central atom. Since we don't yet know how to draw covalent compounds, how do we know which is the central atom? The central atom is usually the one that's least abundant in the compound. For example, the central atom of CF_4 is carbon.

Rule 2

The second word is the name of the other atom in the compound, with "-ide" replacing the end of the element name. For example, at this point we would refer to CF_4 as "carbon fluoride."

Rule 3

Prefixes will sometimes need to be added to the beginning of each word to indicate that more than one atom of the element is present. The most commonly used prefixes are shown in the following table.

Number of Atoms	Prefix
1	mono- (use only for oxygen)
2	di-
3	tri-
4	tetra-
5	penta-
6	hexa-
7	hepta-
8	octa-

In our example of CF_4, carbon doesn't require a prefix (the only time we ever use a prefix for one atom is with oxygen), and fluorine will have the "tetra" prefix. As a result, CF_4 is known as "carbon tetrafluoride."

Rule 4

A few very common molecules have names that don't follow this system. The most important include water (H_2O), ammonia (NH_3), and methane (CH_4). If you name these compounds using the steps above, people will have no idea what you're talking about!

If you are presented with covalent compounds that consist of two or more atoms of the same element bonded together, the name of the molecule is the same as the name of the element. For example, Cl_2 is simply called "chlorine."

You've Got Problems

Problem 1: Name the following covalent compounds:
(a) PCl_3
(b) CO
(c) SF_6

Writing Formulas from Names

Writing formulas from names is the opposite of the process you just learned. For example, if you find that a chemical compound is called "nitrogen trichloride," this suggests that there is one nitrogen atom and three chlorine atoms, giving you a formula of NCl_3. It's not too hard, and you won't have problems with this.

The Mole Says

If you examine the positions of the seven diatomic elements on the periodic table, six of them form the shape of a seven on the right side of the periodic table and hydrogen is left all by itself in the top left. As a result, it's easy to remember the seven diatomic elements as being "the big seven and hydrogen."

The only thing that might give you trouble are some of the elements. When most elements are named, you simply write the atomic symbol of the element. For example, "carbon" is written as C. However, some of the elements are diatomic, meaning that they naturally occur in molecules containing two bonded atoms. These elements include the halogens (F_2, Cl_2, Br_2, I_2), oxygen (O_2), nitrogen (N_2), and hydrogen (H_2). As a result, if anybody tells you that they're doing a reaction with any of these seven elements, you'll need to remember the formulas above.

You've Got Problems

Problem 2: Write the formulas of the following covalent compounds:
(a) hydrogen bromide
(b) silicon dioxide
(c) oxygen dichloride

The Least You Need to Know

- Covalent compounds are formed when two electronegative atoms are forced to share one or more pairs of electrons with one another.
- Valence electrons are the number of s- and p-electrons since the most recent noble gas, and are responsible for the reactivities of most elements.
- Single covalent bonds are created when two atoms share one pair of electrons, while multiple covalent bonds are formed when they share more than one pair.
- The properties of covalent compounds depend strongly on the fact that covalent molecules are not chemically bonded to one another.
- It's not hard to name covalent compounds.

Chapter 10

Bonding and Structure in Covalent Compounds

In This Chapter

- What hybrid orbitals are, and how they're formed
- Learn how to draw Lewis structures
- The valence shell electron pair repulsion (VSEPR) theory

I've got good news and bad news for you. The good news is that you now understand that covalent molecules are formed when two atoms with similar electronegativities react with one another. The bad news is that we don't yet know how this happens.

As it turns out, electronegativity isn't enough to explain the bewildering variety of covalent compounds that exist. It explains why we see covalent bonds, but explains neither the shapes of the molecules that are formed nor the number of bonds they want to form.

In this chapter, we're going to talk about the mysteries of hybrid orbitals, the Valence Shell Electron Pair Repulsion Theory (VSEPR), and Lewis structures. It's not a chapter for the faint of heart, but I have full confidence that you'll get through it without too much trouble.

Covalent Compounds Get Mysterious

Let's recap what we know about how covalent molecules bond:

- When two atoms with similar electronegativities bond with each other, they form covalent compounds.
- Since both atoms want more electrons to be like the nearest noble gas (because of our old friend, the octet rule), neither atom wants to transfer electrons to the other. To form compounds, they need to share electrons instead.
- We also understand a whole bunch of stuff about the properties and naming of covalent compounds.

None of what you've learned so far is wrong. However, your understanding of bonding in covalent compounds isn't yet complete. Specifically, while we know that electrons are shared between two atoms in a covalent molecule, we don't have any ideas about the locations of these electrons. As you might imagine, they're located within orbitals, but what sort of orbitals exist between two atoms?

Why are orbitals important? As it turns out, covalent bonds are formed when two orbitals from different atoms, each of which have one electron in them, overlap so that these two electrons are shared. Because these orbitals need to overlap for a bond to be formed, it's important that we understand the shapes of orbitals that are formed in covalent compounds.

Before we can fully understand the true nature of orbitals in covalent compounds, we must see what's incomplete about our current understanding of orbitals and electrons. Imagine that four hydrogen atoms have bonded with one carbon atom to form methane, CH_4. The type of diagram we saw in Chapter 9 makes this seem like a simple matter.

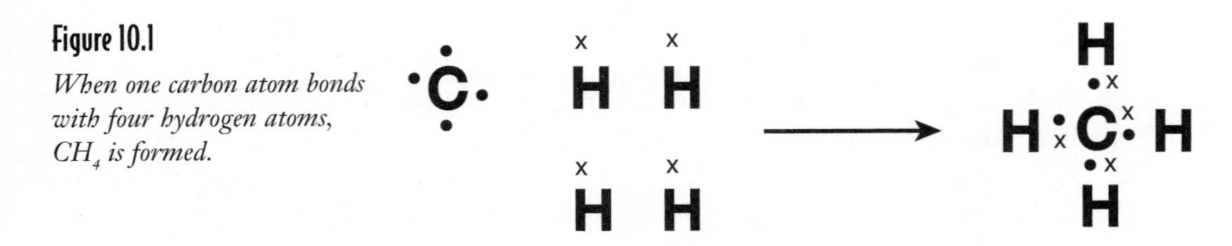

Figure 10.1

When one carbon atom bonds with four hydrogen atoms, CH_4 is formed.

Though this gives us a nice conceptual view of what's going on, it doesn't show us what actually happens to the s- and p-orbitals on carbon when these elements bond to form methane. From this diagram, it looks like the atoms want to be 90° apart from each other. However, when we check what the bond angles in methane really are, we find that they're 109.5°. What are we to do?

The Mystery and Wonder of Hybrid Orbitals

In order for the electrons to spread further than 90° from each other, we have to come up with a new model that allows for this.

Up to now, we've learned the shapes and relative energies of s-, p-, d-, and f-orbitals. However, when atoms form covalent compounds, atomic orbitals are insufficient because they force the bonded atoms to be too close to each other. As you might expect, the electrons in covalent bonds, as is the case with electrons everywhere, prefer to be as far apart from each other as possible because they repel each other. Subsequently, all orbitals within an atom that contain valence electrons combine with one another to form "hybrid orbitals."

It's probably easiest to understand how hybrid orbitals work by showing you an example. Let's take a look at the orbital filling diagram for the valence electrons on carbon (see Chapter 5 for more on orbital filling diagrams).

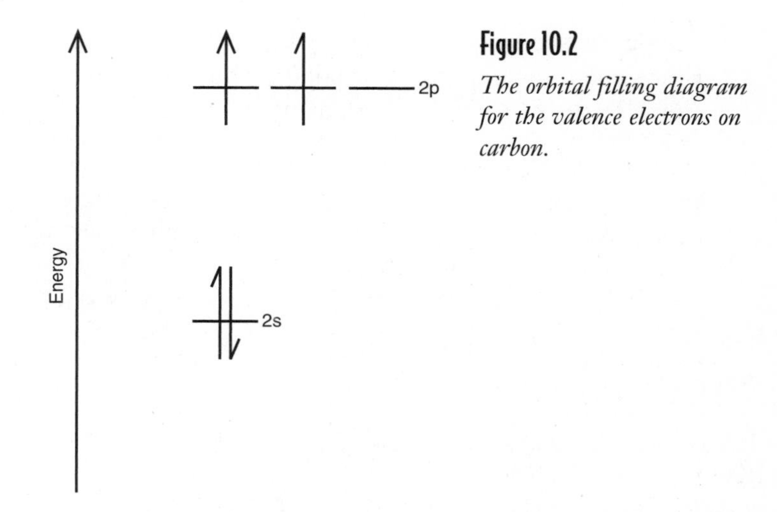

Figure 10.2
The orbital filling diagram for the valence electrons on carbon.

As you can see, of the four valence orbitals in carbon, one is filled (2s), two are half-filled (2p), and one is completely empty. This may be the electron configuration of an unbonded carbon atom, but it doesn't explain how carbon can bond four times to form methane. After all, each covalent bond requires the overlap of an orbital containing one electron from each atom. If this model were valid, we would have no bonding with the s-orbital because it's already full, two bonds (one each from two p-orbitals), and one p-orbital that was completely empty. As a result, carbon could only bond twice, a conclusion that doesn't match reality.

def•i•ni•tion

Hybrid orbitals are formed by mixing two or more of the outermost orbitals in an atom together. The only element that doesn't form hybrid orbitals is hydrogen, as it has only a single 1s orbital.

What really happens when carbon covalently bonds with other elements is that these four dissimilar s- and p-orbitals mix with one another to form four identical hybridized orbitals. The names of these new hybridized orbitals are a combination of the names of the original four atomic orbitals. In our example, one s-orbital combines with three p-orbitals to form four sp^3 orbitals:

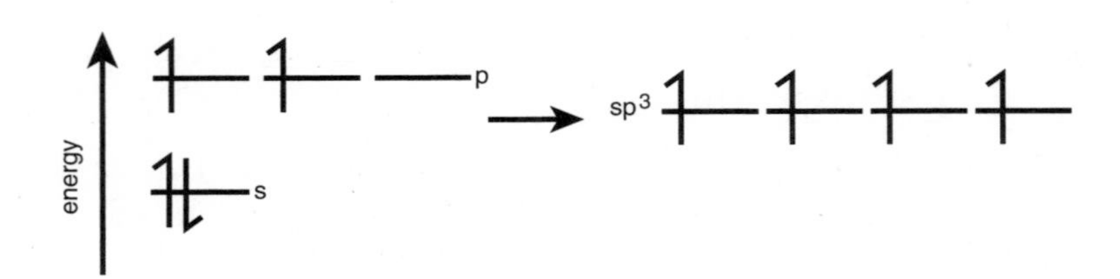

Figure 10.3

When orbitals combine to form hybrid orbitals, both their shapes and energies are averaged.

As you can see from this diagram, the hybridized orbital configuration of carbon allows room for four covalent bonds, which matches well with the four hydrogen atoms covalently bonded in methane.

As we mentioned in Chapter 5, s-orbitals are spherical and p-orbitals are at 90-degree angles to one another. As a result, the sp^3 orbitals that form when one s- and three p-orbitals combine will be arranged at an angle reflecting a mixture between these two types of orbitals (in this case, 109.5°):

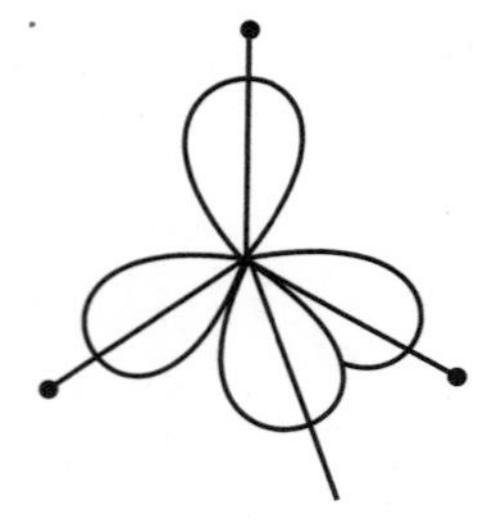

Figure 10.4

As you can see, these four orbitals are pointed away from each other as far as they can go, which minimizes the repulsions between bonding electron pairs.

The number of hybrid orbitals that are formed when a covalent molecule bonds depends on the number of single bonds and pairs of unbonded electrons (lone pairs or unshared electron pairs) that are present in the molecule. The electrons in both single bonds and unbonded pairs exist within hybrid orbitals.

The electrons in multiple bonds exist within something called a "π-orbital" that's formed when an unhybridized p-orbital from one atom overlaps with an unhybridized p-orbital from another atom. Let's see how this works in oxygen, O_2.

The Mole Says

The electrons on unbonded atoms are located in s-, p-, d-, and f-orbitals. The electrons in atoms that have formed covalent compounds exist within hybrid orbitals.

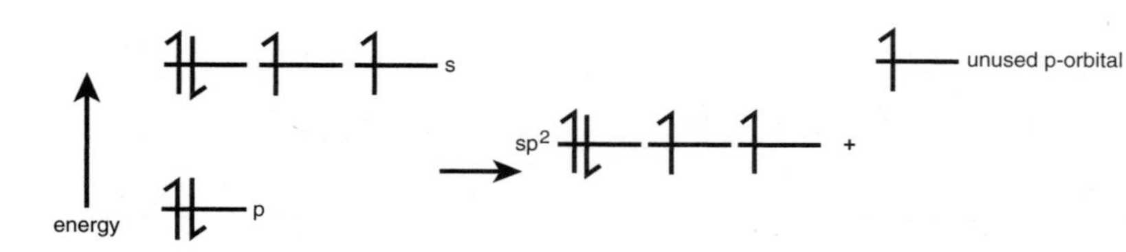

Figure 10.5

In molecular oxygen, only two of the p-orbitals mix with the s-orbital to form three sp² orbitals. The third, unused, p-orbital on each oxygen atom is responsible for the double bond.

In O_2 the first of the two bonds between the two oxygen atoms requires hybrid orbitals (because all single bonds require hybrid orbitals), but the second utilizes the spare p-orbital in both atoms. As a result, one s-orbital mixes with two p-orbitals to form three sp^2 orbitals.

Let's take a look at what the orbitals on each oxygen atom look like.

Figure 10.6

The three sp^2 orbitals, which hold the lone pair of electrons and the electrons shared in single bonds, move as far away from one another as possible, at 120° angles. The p-orbitals on each oxygen atom (not shown) overlap to form the double bond.

The three sp^2 orbitals spread out as far away from one another as possible due to electron repulsion. The p-orbitals, which are responsible for the double bond, overlap both above and below the bond.

The following table illustrates each type of hybrid orbital that commonly exists in covalent compounds, as well as the name and bond angles of each orbital and the names of each molecular shape.

Nonhybridized Overlapping Orbitals	Name of Hybrid Orbital	Bond Angle	Molecular Shape
1s, 1p	sp	180°	linear
1s, 2p	sp^2	120°	trigonal planar
1s, 3p	sp^3	109.5°	tetrahedral
1d, 1s, 3p	dsp^3	90°, 120°	trigonal bipyramidal
2d, 1s, 3p	d^2sp^3	90°	octahedral

In common terms, single covalent bonds between two atoms are referred to as "sigma bonds," or σ bonds. These sigma bonds are created by the overlap of two hybrid orbitals. Each multiple bond is referred to as a "pi bond," or π bond. Pi bonds are created by the overlap of non-hybridized p-orbitals. Using the preceding table, each atom in the sp hybridized atom above has two σ bonds and two π bonds.

Drawing Lewis Structures

Up until this point, we've been determining the types of hybrid orbitals by examining pictures of each molecule being studied. Inevitably, the time has come for you to make some molecular diagrams of your own. These diagrams, called *Lewis structures*, show all of the valence electrons and atoms in a covalently bonded molecule.

def•i•ni•tion

Lewis structures (named for chemical theorist Gilbert Newton Lewis) are pictures that show all of the valence electrons and atoms in a covalently bonded molecule.

If you've been exposed to Lewis structures before, you may have the erroneous idea that they're difficult to draw. The reason for this is simple: It's a difficult concept for teachers to explain, and books don't usually do much better. Fortunately, I have a foolproof method that can make anybody into a Lewis structure king or queen.

Follow these steps.

Step 1

Count the total number of valence electrons in the molecule.

As an example, let's use carbon tetrachloride, CCl_4. The single carbon atom contains four valence electrons, and each of the four chlorine atoms contains seven valence electrons. Therefore, the number of valence electrons for this molecule is $4 + (4 \times 7) = 32$.

Occasionally, you'll have to find the Lewis structure for a polyatomic ion. To do so, do everything normally as mentioned above. When you're done, you'll need to adjust this number to account for the ionic charge. For cations, you'll have to subtract the charge and for anions you'll have to add the charge. For example, you'll subtract one electron from the count for NH_4^+ and add two electrons to the count for CO_3^{-2}.

Step 2

Count the total octet electron count in the molecule.

You won't find the term "octet electron count" in any textbook because, as far as I know, I made it up. The number of "octet electrons" is equal to the number of valence electrons that each atom will have when they have the same electron configuration as the nearest noble gas (the octet rule). The number of octet electrons that atoms want can usually be determined by the following rules.

- Hydrogen wants two octet electrons.
- Boron wants six octet electrons for neutral molecules and eight for molecules with charge.
- All other atoms want eight octet electrons.

In our example, carbon wants eight octet electrons and each of the four chlorine atoms also want eight octet electrons. The total number of octet electrons for the molecule will then be equal to $8 + (4 \times 8) = 40$.

Step 3

Subtract the number of valence electrons from the number of octet electrons to find the number of electrons that are involved in bonding.

In our example, $40 - 32 = 8$ bonding electrons.

Bad Reactions

If you find that you have a fraction in your number of bonds, you've made a mistake in an earlier step (usually step 1). Go back to the beginning and check your work!

Step 4

Divide the number of bonding electrons by two to find the number of bonds.

Because there are two shared electrons in every covalent bond, dividing the bonding electrons by two can be used to find the number of chemical bonds. In our example, 8/2 = 4 bonds.

Step 5

Arrange the atoms so the molecule has the same number of covalent bonds that you found in step 4.

In this step, it's tempting to just randomly stick atoms and bonds wherever you can until everything is stuck together. Unfortunately, randomness rarely yields the right answer, so we'll need some rules to help us out.

- The atom that's least abundant in the compound is usually in the middle of the molecule. In our example, we can assume that carbon will probably be in the center of the molecule.
- Hydrogen and the halogens bond once.
- Oxygen's family bonds twice in uncharged molecules and one, two, or three times when present in polyatomic ions.
- Nitrogen's family bonds three times in uncharged molecules and two, three, or four times when present in polyatomic ions.
- Carbon's family usually bonds four times.
- Boron usually bonds three times in uncharged molecules and four times when present in a polyatomic ion.

In our example, CCl_4, there are fewer carbon atoms than chlorine atoms, so the carbon atom goes into the middle of the molecule, with four chlorine atoms arranged around it. Between the carbon and each chlorine atom is a single chemical bond, totaling four. In this structure, both carbon and chlorine follow the rules for the number of bonds each wants. Even better, we've used the number of bonds we thought we'd need from step 4!

Cl
|
Cl – C – Cl
|
Cl

Figure 10.7

So far, so good!

Step 6

Add lone pairs of electrons to each atom until each atom is surrounded by the number of electrons we said they wanted in step 2.

Let's take a look at the carbon atom in our diagram. The four bonds around it contain eight electrons. Since carbon wants eight electrons, it doesn't require lone pairs.

Each chlorine, on the other hand, has only one bond for a total of two electrons. Since chlorine wants eight electrons, three pairs need to be added to each. The completed Lewis structure for CCl_4 is shown in the following figure.

The Mole Says

If you place single bonds between all of the atoms and there are still some left over, you may need to start putting in double or triple bonds. There's nothing wrong with this—just make sure that all of the atoms have the correct number of bonds when you're done!

:Cl:
|
:Cl – C – Cl:
|
:Cl:

Figure 10.8

The final Lewis structure of carbon tetrachloride. I told you it wasn't that hard!

Step 7

Check to see if any of the atoms in the molecule have a positive or negative charge.

Add the number of lone pair electrons to the number of bonds for each atom in the molecule. Subtract this number from the number of valence electrons the atom is expected to have (which you already figured out in Step 1) to calculate its charge. Once you've determined the charge on an atom, write it next to the atomic symbol of that atom.

In our example, CCl_4, carbon has four bonds around it. Because it normally has four valence electrons, carbon has no charge (4 – 4 = 0). Because each chlorine has one bond and six lone pair electrons (1 + 6 = 7) in this diagram and seven valence electrons, it also has no charge.

You've Got Problems

Problem 1: Draw the Lewis structures for the following molecules or polyatomic ions, showing the charges on each atom (if needed).

(a) NH_3

(b) SiO_2

(c) OH^{-1}

Resonance Structures

For an additional practice problem, try drawing the Lewis structure for the nitrate ion, NO_3^{-1}. This explanation will make a lot more sense if you really draw it, so get a sheet of paper and do it now. I'm going to go get a snack while you work on this.

When you drew the Lewis structure for the nitrate ion, you (hopefully) came up with one of these three structures.

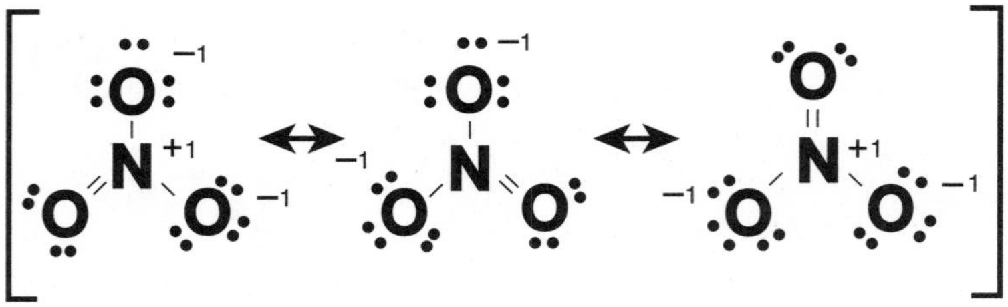

Figure 10.9

These represent the three equivalent resonance structures for the nitrate ion.

These three figures are the resonance structures of the nitrate ion. When more than one valid Lewis structure can be drawn for a given arrangement of atoms in a covalent compound, they are referred to as *resonance structures*. In resonance structures, all of the atoms are located in exactly the same positions, but the numbers and/or locations of the electrons (bonds or lone pairs) may be different.

def•i•ni•tion

Resonance structures occur when more than one valid Lewis structure can be drawn for a given arrangement of atoms in a covalent compound. In resonance structures, the atoms are all in the same positions, but the number and locations of bonds and lone pair electrons may be different. The true form of the molecule is an average of the resonance structures that can be written for it.

You may be wondering which of the three equivalent resonance structures is the true structure of the nitrate ion. As it turns out, the actual structure of this ion is an average of the three. Instead of one double bond and two single bonds between the nitrogen and three oxygen atoms, imagine a situation where there's actually $1\frac{1}{3}$ bonds between each of the atoms. Likewise, each oxygen atom really has a $-\frac{1}{3}$ charge. Because the concept of odd numbers of bonds and uneven charge is confusing, we usually just draw all of the possible resonance structures for a molecule and let it go at that.

You've Got Problems

Problem 2: Draw both of the possible resonance structures for the formate ion (CHO_2^{-1}). Remember to show the charges on each atom!

Valence Shell Electron Pair Repulsion Theory (VSEPR)

Take a look at the heading for this section. *Valence Shell Electron Pair Repulsion Theory (VSEPR)*. Say it five times quickly. No matter how you look at this phrase, it doesn't sound like something fun.

However, I've got great news for you. The VSEPR (pronounced "vesper") theory is something we've already discussed. VSEPR theory simply states that the pairs of electrons in a chemical compound repel each other and move as far from each other as possible because they have the same charge.

See, I told you it wasn't that hard. We already knew electrons repel each other—heck, we already talked about it a little bit earlier in this chapter. The big question, of course, is "how will this theory affect us personally?" Good question.

def•i•ni•tion

Valence Shell Electron Pair Repulsion Theory (VSEPR) states that the shapes of covalent molecules depend on the fact that pairs of valence electrons repel each other and move as far away from each other as possible.

What Does VSEPR Have to Do with Hybrid Orbitals?

When we discussed hybrid orbitals, we mentioned the bond angles associated with each type of orbital. These bond angles represent the greatest possible angles between neighboring pairs of electrons. For example, if we look at a picture of the three sp^2 orbitals in an atom, we can see that they stretch as far away as possible, 120°.

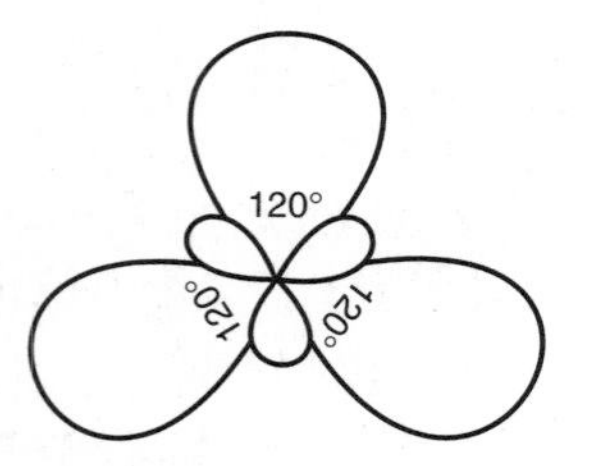

Figure 10.10

The farthest that the electrons in the three bonds can be distanced from each other in space is 120°. The shape associated with this bond angle is "trigonal planar."

As it turns out, there's a little more to knowing bond angles than just knowing the hybridization of the central atom in a molecule. While the pairs of electrons in a covalent bond are situated mainly between the two bonding atoms, the electrons in a lone pair spread out a little bit. Because these lone pairs spread out their negative charge, the chemical bonds are repelled a bit more by lone pairs than by each other. Therefore, the bonds in covalent compounds with lone pairs tend to be squished together and have smaller bond angles than predicted solely by the hybridization. This phenomenon is shown in the following figure.

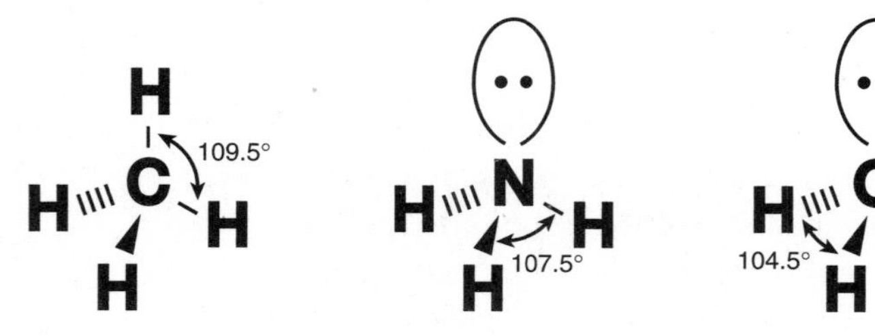

Figure 10.11

Though all three of these molecules exhibit sp^3 hybridization around the central atom, the bond angles get successively smaller because of the increasing numbers of lone pairs.

Bond Angles and Molecular Shapes the Easy Way

So how can ordinary people like you or me remember the bond angles and shapes of all the atoms in all the covalent compounds known to man? We could memorize them, but that would be really boring and cut into our valuable television time. Instead, we'll use the Lewis structures we learned earlier to give us a hint about how covalent molecules are put together.

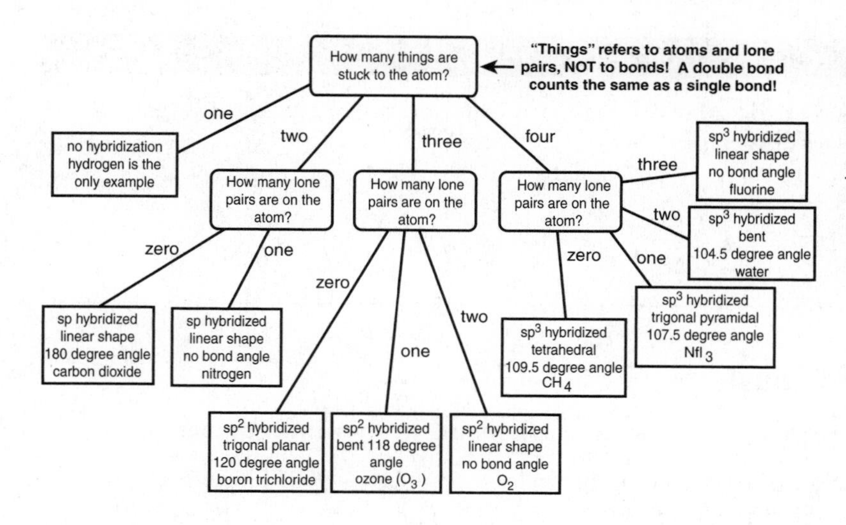

Figure 10.12

Once you have a valid Lewis structure, you can use this flow chart to find the hybridization, bond angle, and shape of any atom on any covalent compound!

Here's how you use this flow chart.

- At the top, you're asked, "How many atoms are bonded to the atom of interest?" In our case, the atom of interest is whatever atom you're trying to find the hybridization and shape of. For PBr_3 (see the following figure), the answer would be "3." Follow the arrow labeled "3" to the next question.

:Br – P – Br:
|
:Br:

Figure 10.13

The Lewis structure of phosphorus tribromide.

- The second question asks, "How many lone pairs does the atom of interest have?" In our case, phosphorus has one lone pair. As a result, we follow the arrow marked "1" to find that phosphorus tribromide is trigonal pyramidal, has a bond angle of 107.5°, and is sp^3 hybridized.

This flow chart can be used to find the hybridization, bond angle, and shape of any covalently bonded atom. The next time you're at a dinner party, you can use this information to wow the guests with your immense knowledge of hybridization and VSEPR theory.

You've Got Problems

Problem 3: Find the hybridizations, shapes, and expected bond angles of the central atoms in the following molecules:

(a) SiO_2

(b) CH_2O

(c) OF_2

The Least You Need to Know

- Hybrid orbitals are necessary to explain bonding in covalent compounds.
- Lewis structures can be drawn by following seven easy steps. These diagrams show us where all of the atoms and valence electrons in a molecule are.
- When there is more than one possible correct way of drawing the Lewis structure of a compound, these drawings are called resonance structures. The actual structure of a molecule is an average of its resonance structures.
- The shapes of covalent molecules are described by the valence shell electron pair repulsion (VSEPR) theory, which states that the electrons in covalent molecules tend to want to spread as far away from each other as possible.

Chapter 11

The Mole

In This Chapter

- The mole defined
- Molar mass
- Moles, molecules, and mass calculations
- Mass percent problems

Depending on your educational background and life experiences, different things may come to mind when the word "mole" is mentioned.

If you've studied the life sciences, you probably think of moles as members of the species *Talpa europaea*, a small burrowing animal. If you work for the CIA, you probably think a mole is somebody who has infiltrated your ranks to steal state secrets. If you've spent a lot of time outside without wearing sunscreen, a mole is a dark-colored, raised skin blemish that sometimes grows big weird hairs.

All of these definitions are true and work well for various purposes. However, moles are none of these things in a chemistry class. In this chapter, we'll examine the mysterious mole and find out what it's good for.

What's a Mole?

I'm wearing a pair of shoes right now. My question for you: How many shoes am I wearing? If you're thinking "two," you obviously know your footwear.

Another pop quiz: I'm feeling ill because I just ate a dozen eggs. How many eggs did I eat? If you said "12," you're clearly in tune with the poultry industry.

One more pop quiz: If I bought a ream of paper, how many sheets of paper did I purchase? If you said "500," you're ready to become a world-class paper salesman!

So what does all this have to do with the mole? As you're already aware, atoms and molecules are very, very small. As a result, it doesn't make much sense to count them individually when doing a chemical reaction. As a result, scientists have come up with a shorthand term for a very large number of molecules, just as shoe salesmen have a term for two shoes, grocers have a term for 12 eggs, and paper salesmen have a term for 500 sheets of paper. That term is *mole*, and it stands for $6.022136736 \times 10^{23}$ things (we'll usually just round it to 6.02×10^{23}).

def•i•ni•tion

A **mole** is equal to 6.02×10^{23} things. Though you could, in theory, have a mole of anything, this number is so huge that we usually only speak of having moles of atoms or molecules, because both are very tiny.

Chemistrivia

6.02×10^{23} is usually referred to as Avogadro's number and is named after the chemist Amadeo Avogadro, who helped to understand the nature of gases.

In our everyday lives, moles aren't a particularly handy unit for measuring numbers of things. For example, let's say that the publisher of this book decided to print a mole of copies for sale worldwide. Such a number of books would require a warehouse with a volume of 512 billion cubic kilometers. Unfortunately, my agent wasn't able to convince the publisher that this would be a good business move. As a result, we should only use "moles" to describe numbers of really small things like atoms or molecules.

Molar Mass

When speaking about doing chemical reactions, it isn't very handy to say that you want "one mole of compound X to react with two moles of compound Y." That may be what you need to do, but unfortunately, there are no machines that can count 6.02×10^{23} molecules.

As a result, we're going to have to learn to find something called the *molar mass*—the weight of one mole of a chemical compound. That way, if somebody says that we need to use two moles of water in a reaction, we can go to a balance and just weigh it instead of having to individually count a very large number of molecules.

def•i•ni•tion

The **molar mass** of a substance is the weight of 6.02×10^{23} atoms or molecules of that material in grams. The unit of molar mass is given as grams/mole, usually abbreviated as g/mol.

Finding the molar masses of compounds isn't difficult if you know their formulas. To do this, multiply the numbers of atoms of each element in a compound by their atomic masses from the periodic table. When you add these numbers up, you come up with the molar mass of the compound.

Chemistrivia

Other common terms that mean the same thing as "molar mass" are "molecular mass," "molecular weight," and even "gram formula mass."

For example, let's find the molar mass of sulfuric acid, H_2SO_4:

Element	Number of Atoms	Atomic Mass (g)	Mass × Atoms
H	2	1.01	2.02
S	1	32.07	32.07
O	4	16.00	64.00
		Total:	98.09 g

As a result, we would say that the molar mass of sulfuric acid is "98.09 g/mol."

You've Got Problems

Problem 1: Find the molar masses of the following compounds:

(a) Na_2SO_4

(b) nitrogen trichloride

(c) fluorine

Converting Between Moles, Molecules, and Grams

Once you know how to find molar mass, you can start to convert between moles, grams, and molecules of a substance. To do so, use the following figure:

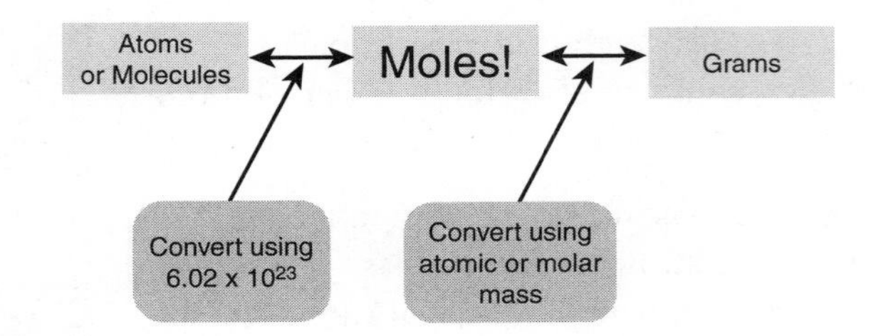

Figure 11.1

By using this map as a guide, you can easily learn to convert between grams, molecules, and moles of chemical compounds.

The method you'll use to solve this problem is the factor-label method, explained in Chapter 2. We'll see how to use the factor label method using the following example:

Example: Convert 63 grams of ammonia to molecules.

Solution: To answer this problem, do the following steps:

1. Write down the number you're trying to convert, along with the units that were given to you.

   $63 \text{ g } NH_3$

2. Write a multiplication sign after the unit you're trying to convert, followed by a straight, horizontal line.

   $63 \text{ g } NH_3 \times \underline{\qquad\qquad}$

3. In the space below the line, write the unit of the number you're trying to convert.

   $$63 \text{ g } NH_3 \times \frac{\qquad\qquad}{\text{g } NH_3}$$

4. Take a look at the map in the preceding figure. Find your starting point (in this case, "g") and move one box over toward your destination. Write the unit that you find in the next box (in this case, "mol") on top of the line. Then, put the formula of the compound you're converting (in our case, NH_3) after the unit.

   $$63 \text{ g } NH_3 \times \frac{\text{mol } NH_3}{\text{g } NH_3}$$

5. Write the conversion factor in front of the units you wrote above and below the line. This conversion factor can be found on the line between the starting and destination boxes in the figure (in this case, it's the molar mass of NH_3).

$$63 \text{ g } NH_3 \times \frac{1 \text{ mol } NH_3}{17 \text{ g } NH_3}$$

6. Multiply these numbers together, making sure to cancel out any appropriate units.

$$63 \cancel{\text{g } NH_3} \times \frac{1 \text{ mol } NH_3}{17 \cancel{\text{g } NH_3}} = 3.7 \text{ mol } NH_3$$

7. If you haven't yet finished the problem, use the output from your first calculation to start a second calculation using steps 1–6.

Bad Reactions

When using this method, always write "1" in front of the word "moles," "6.02×10^{23}" in front of the unit "atoms" or "molecules," and the molar mass in front of the unit "grams." If you don't, your answer will be wrong!

In this case, we found that we have 3.7 moles of NH_3. However, we're trying to find the number of molecules of ammonia, not the number of moles. As a result, we need to go through the preceding six steps to find molecules. When this calculation is set up, it should look like this:

You've Got Problems

Problem 2:

(a) How many grams are there in 4.3×10^{22} molecules of $POCl_3$?

(b) How many moles are there in 23 grams of sodium carbonate?

$$3.7 \cancel{\text{mol } NH_3} \times \frac{6.02 \times 10^{23} \text{ molecules } NH_3}{1 \cancel{\text{mol } NH_3}} = 2.2 \times 10^{24} \text{ molecules}$$

Our answer, then, is 2.2×10^{24} molecules NH_3.

Percent Composition

Sometimes, it's handy to figure out how much of an element is present in a chemical compound. Let's say that you're trying to add extra calcium to your diet by taking a calcium supplement containing 1 gram of $CaCO_3$. How much calcium are you actually getting?

To solve this problem, we need to figure out the percentage of calcium present in this compound. This percentage is referred to as either the mass percent or weight percent and is found by using the following formula:

$$\text{Mass percent} = \frac{\text{mass of the element we're interested in}}{\text{molar mass of the whole compound}} \times 100\%$$

Let's use this equation to find the amount of calcium in one gram of a $CaCO_3$ supplement:

The mass of calcium in this compound is equal to 40.1 grams because there's one atom of calcium present and calcium has an atomic mass of 40.1 amu. The molar mass of the compound is 100.1 grams. Using the handy equation above, we get:

$$\text{Mass percent} = \frac{40.1 \text{ g Ca}}{100.1 \text{ g } CaCO_3} \times 100\% = 40.1\% \text{ Ca}$$

You've Got Problems

Problem 3:

(a) Find the percent composition of silver in silver nitrate.

(b) How many grams of silver are present in 25 grams of silver nitrate?

Our result means that 40.1% of the mass in the calcium carbonate supplement is caused by calcium. To find the total mass of calcium in the supplement, we find 40.1% of one gram to get a final answer of 0.401 grams of calcium. Likewise, if we wanted to find the amount of calcium in 125 grams of the supplement, we'd find 40.1% of 125 grams, or 50.1 grams.

The Least You Need to Know

- A mole is equal to 6.02×10^{23} things. We use it only for atoms and molecules because both are very small.
- It's possible to convert between grams, moles, and molecules of a compound using the factor-label method.
- The mass percent of a compound is important in determining how much of each element is present.

Part 3

Solids, Liquids, and Gases

As you've probably already noticed, atoms and molecules are really, really small. As a result, it's hard to figure out what they're doing. Are they sitting quietly on your desk or plotting to overthrow the government? You can't tell, because they're too small to see.

In Part 3, we'll learn about what atoms and molecules are really thinking. Whether they're uptight and rigid solids, squishy and sloshy liquids, or hyperactive and speedy gases, we'll figure out not only what they look like, but also what sorts of things they're likely to do.

Chapter 12

Solids

In This Chapter

- Descriptions and definitions of solids
- Crystals and crystal structures explained
- The six main types of solids and their properties

You may have noticed that this book assumes you know what solids, liquids, and gases are. Though most chemistry books define these terms early in the first chapter, I think you probably know enough to realize that if I were to throw something solid at your head, it would hurt. Likewise, if I throw a liquid at your head, you will get wet, and blowing air on you will cause you to feel a mild breeze.

However, if you've learned nothing else, you should be aware that chemistry doesn't just deal with macroscopic (big) properties of objects but also with microscopic (small) properties. As a result, it's not enough to know that hitting you in the head with a solid rock will hurt—we want to find out why the rock feels hard and how the atoms in the rock will feel about their impact with your head. In this chapter, we'll learn why rocks are hard, among other things.

What Are Solids?

Solids are the state of matter in which the atoms or molecules are locked into place by either chemical bonds or forces between molecules called "intermolecular forces." Solids are usually hard, have a shape that doesn't change, and possess a fixed volume.

Why Are Solids Solid?

Until you started studying chemistry, you probably took it for granted that it's a bad idea to hit yourself on the head with a rock because solids are hard. Now that your chemical studies have begun, the question that arises is "Why are solids hard in the first place?" To answer this question, we need to examine the nature of crystals.

def•i•ni•tion

Solids are the state of matter in which the atoms or molecules are locked rigidly in place by bonds or intermolecular forces.

def•i•ni•tion

Crystals are regular arrangements of atoms, ions, or molecules stacked into repeating three-dimensional structures. The smallest unit that can be stacked together to re-create the crystal is referred to as a **unit cell**.

Some Basic Definitions of Crystals

The atoms in many solids are locked into rigid groups called *crystals*. The atoms, ions, or molecules in crystals are held together by attractive forces between ions of opposite charge, covalent bonds, and other forces between covalent molecules referred to as intermolecular forces (see Chapter 13). Overall, the three-dimensional structure of a crystal is referred to as a crystal lattice.

Because crystals have regular arrangements, we can think of them as being large structures consisting of building blocks that repeat over and over again. The smallest unit of crystals that can be stacked together to re-create the entire crystal is referred to as a *unit cell*.

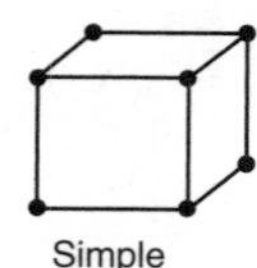

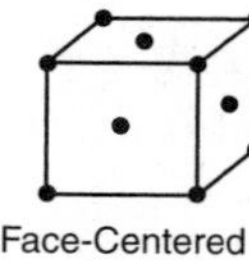

Figure 12.1

These three crystals represent three very common cubic crystal structures—simple cubic, face-centered cubic, and body-centered cubic. The name "cubic" comes from the shape of the unit cell for each crystal.

There are many different types of crystal structures. The type of crystal structure employed by crystalline solids depends on the specific atoms, ions, or molecules present in the crystal. The following figure shows several different types of crystal structures; keep in mind that these are only a sampling of the different types of crystals and are not meant to represent all of the possible types of crystal structures.

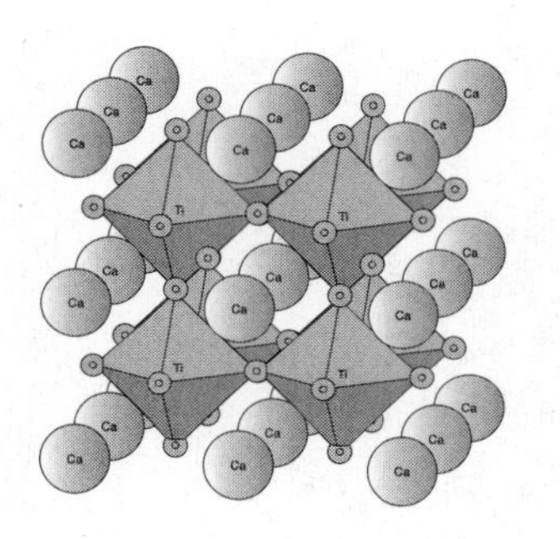

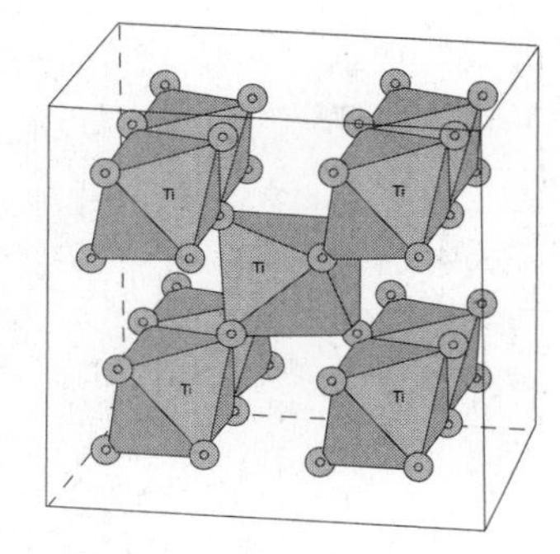

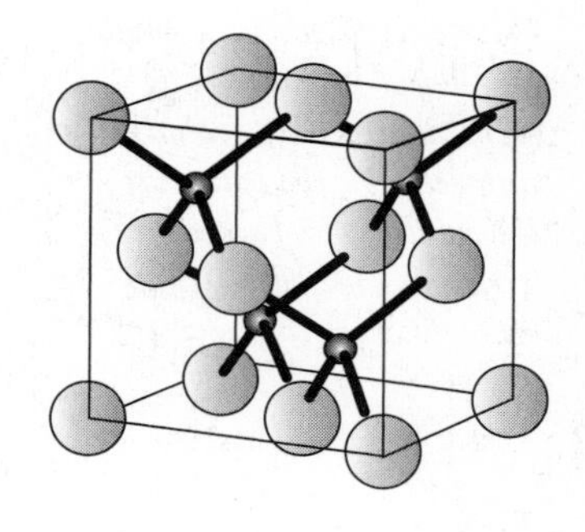

Figure 12.2

From left to right: the crystal structures of $CaTiO_3$, TiO_2, and ZnS.

Close-Packed Crystal Structures

When the atoms in a crystal are all of identical size, they come together in a close-packed structure. As you might guess, the atoms in a crystal are generally only the same size when the atoms are all of the same element. Metals in particular tend to bond in *close-packed* structures.

Close-packed structures are crystals in which each atom is as close to all its neighboring atoms as possible. If you place a bunch of marbles into a glass, they will arrange themselves so that they take up the least possible space—in short, in a close-packed configuration.

There are two different ways that atoms can stack together in a close-packed arrangement. The first way is called the hexagonally close-packed (hcp) arrangement, in which the layers of atoms that make up the crystal alternate in an ABAB pattern, with every third layer being directly above the first layer. The second type of arrangement is called the cubic closest-

def•i•ni•tion

The term **close-packed** refers to a crystal structure in which all of the atoms are as close together as possible. The two types of close-packed structures are the hexagonal closest-packed (hcp) structure and the cubic closest-packed (ccp) structure.

packed (ccp) arrangement. In a ccp arrangement, the layers of atoms are arranged in an ABCABC pattern, allowing every fourth layer to be located directly above the first layer. Both types of packing arrangements are shown in the following figure.

Figure 12.3

On the top is the hexagonal close-packed (hcp) arrangement, with its hexagonal unit cell. On the bottom is the cubic closest-packed (ccp) arrangement, with its face-centered cubic (fcc) unit cell.

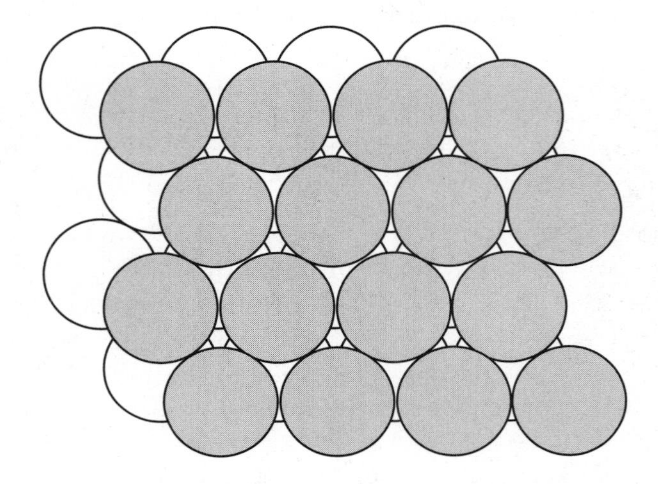

In this crystal, the third row of atoms goes on top of the first (bottommost) one.

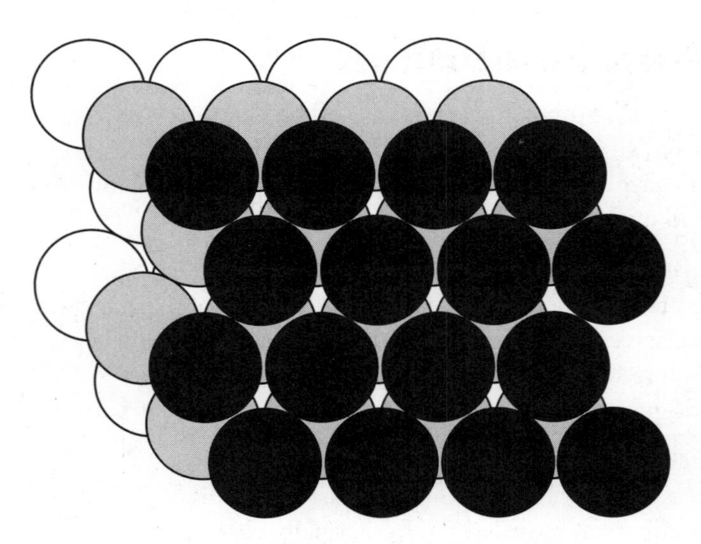

In this crystal, the third row of atoms doesn't go directly on top of the first.

The atoms in metals usually arrange themselves in either an hcp or ccp arrangement. Generally, the more valence electrons a metal has, the more likely it is to have the ccp structure. Some metals take on body-centered cubic or even simple cubic structures, particularly at high temperatures.

Types of Solids

There are five main types of solids, each of which has its own properties and structures. Let's have a look.

Ionic Solids

As we discussed at great length in Chapter 8, ionic solids consist of cations and anions held together by the strength of their opposite charges. The force that holds oppositely charged particles together is called an "electrostatic force."

In Chapter 8, we treated all ionic solids as if they consisted of crystals in which all the ions had identical sizes. As you can probably guess, ions come in a wide variety of sizes. For example, in sodium chloride, the negatively charged chloride ions are much larger than the positively charged sodium ions. As a result, the structure of sodium chloride isn't a simple cubic structure, but rather a face-centered cubic structure. Likewise, the crystal structures of ionic compounds depend on the ratio of the sizes of the anion and cation.

Metallic Solids

Back in Chapter 7 it was explained that metals are good conductors of electricity and heat, have high malleability (bendability), high ductility (can be made into wires), and are shiny. What we never explained was why metals exhibit these properties. As it turns out, the properties of metals stem from the nature of metallic bonds.

One very simple model used to explain bonding in metals is referred to as the electron sea theory. In the electron sea theory, the cations in a metallic solid remain in stationary crystalline positions while the valence electrons from each metal are free to wander throughout the entire solid.

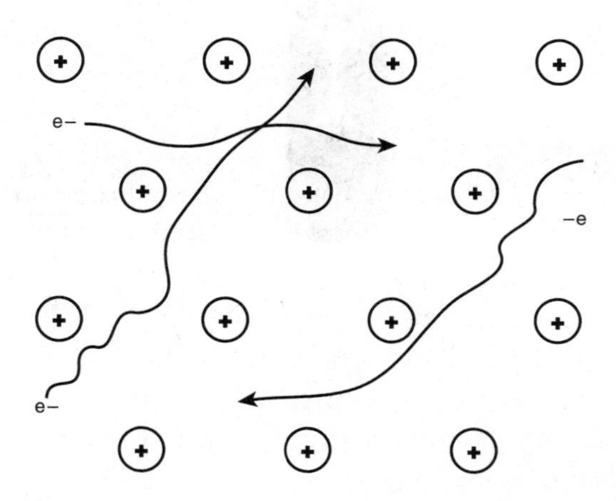

Figure 12.4

In the electron sea theory, the metal nuclei are locked in place and the electrons move freely through the solid.

This theory does a good job of explaining the properties of metals. Because electrons are able to move freely throughout the entire solid, metals are excellent conductors of electricity. The high mobility of electrons also causes metals to conduct heat because they do a good job of dispersing energy. Because metal nuclei can move from place to place without causing bonds to be broken, metals are both malleable and ductile.

Though the electron sea theory accurately describes the properties of metals, it glosses over how the electrons are able to wander freely throughout the solid. After all, didn't we spend a great deal of time learning about how electrons exist within orbitals?

Think back to when we discussed hybrid orbitals (Chapter 10). When an s-orbital and three p-orbitals overlap, they form four sp^3 orbitals. If an s-orbital and two p-orbitals overlap, they form three sp^2 orbitals.

In a metal, a similar thing happens. However, unlike covalent compounds, where only a few orbitals mix, all of the metal atoms mix their atomic orbitals (s-, p-, and d-orbitals) together to form a huge number of orbitals known as "molecular orbitals." These molecular orbitals are similar in energy to one another and form something called a conduction band.

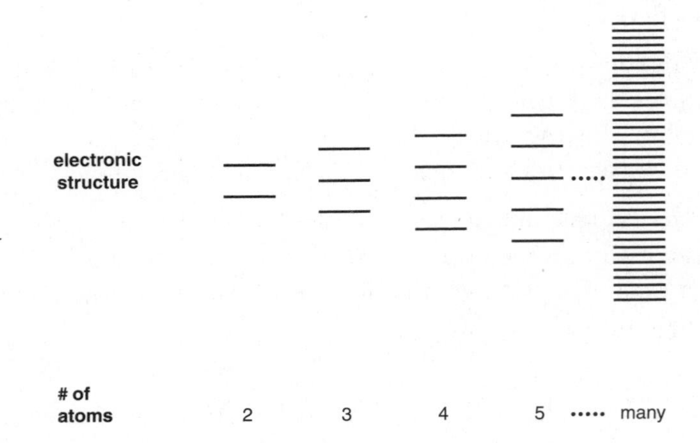

Figure 12.5

As the number of metal atoms increases, the number of molecular orbitals increases. Eventually, these orbitals become so close in energy that they form a conducting band in which electrons can easily jump from one orbital to another.

Chemistrivia

The molecular orbital theory that explains electron motion in metals is also referred to as "band theory."

Because there are more molecular orbitals in the conduction band than there are electrons, it doesn't take much energy to raise an electron from a filled orbital to an empty one with higher energy. When these electrons jump to empty orbitals, they are able to move freely around the metal.

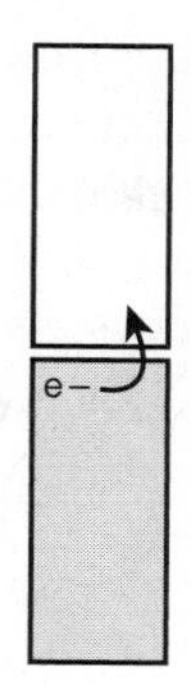

Figure 12.6

In a metal, the difference in energy between filled orbitals and unfilled orbitals is very small. As a result, it's easy for electrons to jump to empty orbitals where they are free to move around the solid.

There are typically several conducting bands in metals. One band (called the "s-band") is caused by an overlap between all of the s-orbitals in the metal. The other bands are called the "p-band" and "d-band" because they result from the overlap of p- and d-orbitals, respectively. Because these bands overlap in energy, they behave as one large, partially filled band.

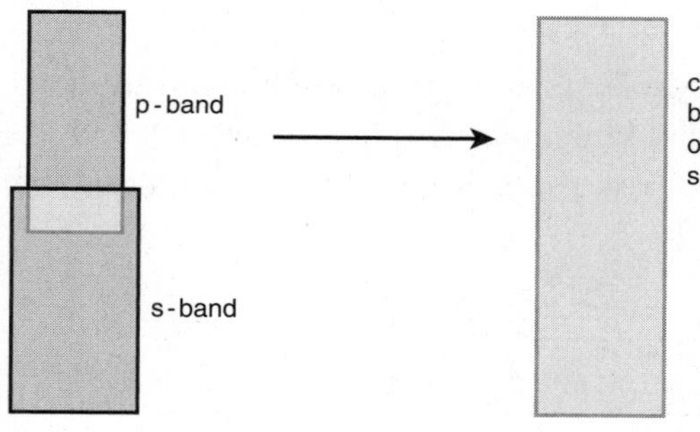

Figure 12.7

When the s- and p-bands overlap in energy, they form a much larger band that makes it possible for metals to conduct electricity.

Other elements are sometimes added to metals to give them desired properties such as hardness, durability, or strength. The resulting material is referred to as an *alloy*.

def•i•ni•tion

An **alloy** is a metallic material in which several elements are present. The elements added to a pure metal to form an alloy are selected to maximize a desired property.

There are two types of alloys:

- Substitutional alloys form when one of the atoms in a metal is replaced with a different element. For example, in sterling silver, some of the silver atoms are replaced with copper.

- Interstitial alloys form when some of the spaces between the atoms in a metal are filled by smaller atoms. One of the most important interstitial alloys is carbon steel, in which carbon atoms are located between iron atoms.

These two types of alloys are shown in the following figure.

Figure 12.8

The picture on the left represents a substitutional alloy; the picture on the right represents an interstitial alloy.

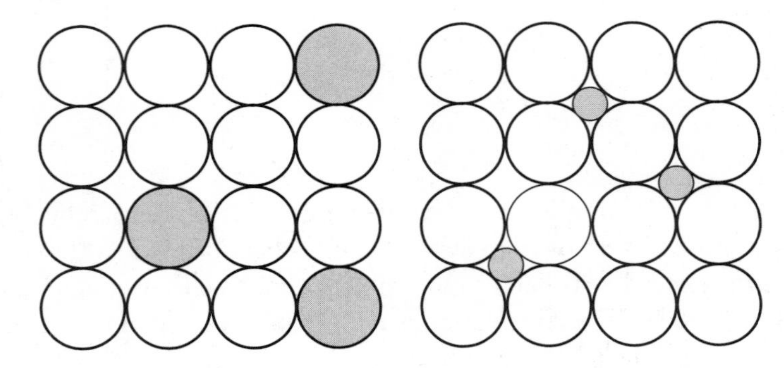

Network Atomic Solids

Network atomic solids are formed when many atoms are bonded together covalently to form one gigantic molecule. Unlike regular covalent molecules that are generally small, network atomic solids may grow quite large. One common example of a network atomic solid is a diamond:

Figure 12.9

The carbon atoms in a diamond are all held together by covalent bonds. As a result, diamonds can be thought of as being very large covalent molecules.

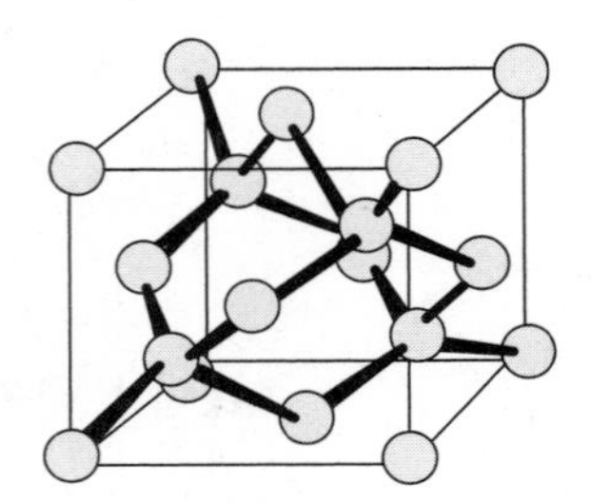

Network atomic solids have a wide number of varying properties. They are usually hard, owing to the strong bonds between neighboring atoms—for example, diamonds are (currently) the hardest known material. They also tend to have high melting and boiling points due to the very strong covalent bonds. Network atomic solids are frequently brittle because a small movement of atoms in the crystal tends to disrupt the network of covalent bonding. Aside from diamonds, some common network atomic solids are quartz (SiO_2), graphite, and silicon.

When elements form network atomic solids, their atomic orbitals (s- and p-orbitals) overlap to form conducting bands in the same way that metals do. However, there is one major difference in the structure of the bands between metals, nonmetals, and metalloids. Whereas the s- and p-bands overlap in metals to form a giant conducting band, they don't for either nonmetals or metalloids, making it difficult for either to conduct electricity.

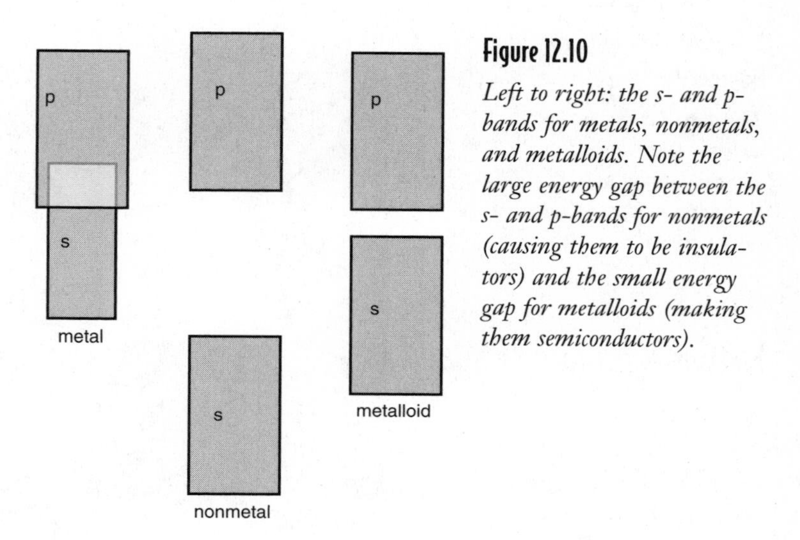

Figure 12.10

Left to right: the s- and p-bands for metals, nonmetals, and metalloids. Note the large energy gap between the s- and p-bands for nonmetals (causing them to be insulators) and the small energy gap for metalloids (making them semiconductors).

Because the s- and p-bands overlap in metals, they are good conductors of electricity. In nonmetals, there is a large energy gap between these bands (called, straightforwardly enough, the "band gap"), making it difficult for electrons to travel from one to another—as a result, they are good insulators. In metalloids, the band gap is small, which means that it is possible for electrons to jump from one band to another, causing them to conduct partially. This explains why metalloids are called "semiconductors."

The Mole Says

A few network atomic solids, such as graphite, conduct electricity. However, these are much less common than insulating network atomic solids.

Molecular Solids

So far, we've talked about solids that are held together by chemical bonds. However, what happens when we make a solid from small covalent compounds, such as water, to form ice?

As it turns out, covalent molecules interact with one another through forces referred to as "intermolecular forces." Though we will spend more time speaking about intermolecular forces in Chapter 13, you can think of them as being Scotch tape in a world of Crazy Glue covalent bonds and ionic attractions. Though intermolecular forces don't possess nearly the strength of ionic attractions or covalent bonds, they are still strong enough to hold covalent molecules together in a solid. One example of a molecular solid is ice.

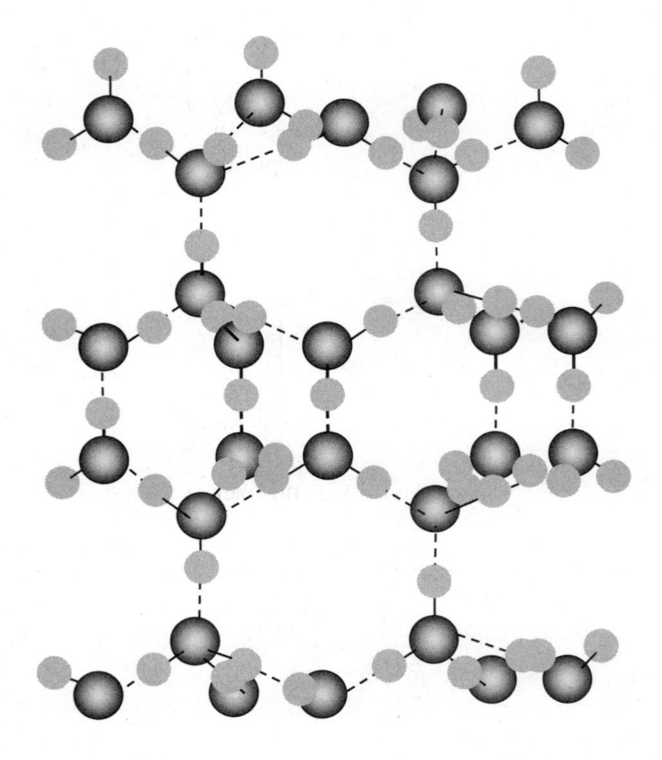

Figure 12.11

The dotted lines in this structure correspond to the intermolecular forces holding the water molecules together in the crystal.

Intermolecular forces are weaker than chemical bonds, making molecular crystals less tightly held together than other forms of crystals. As a result, molecular solids frequently have low melting points and are easily broken apart. Molecular solids are also extremely poor conductors of electricity. Aside from ice, other examples of molecular solids are sugar and dry ice.

Atomic Solids

Atomic solids are formed when the noble gases become cold enough to freeze. As with molecular solids, there are very weak intermolecular forces known as "London dispersion forces" that hold these atoms together. Because their interactions are extremely weak, frozen noble gases tend to be soft and have very low melting points. (See Chapter 13 for more about London dispersion forces.)

Amorphous Solids

Some solids don't have a particular structure. Instead of being arranged into a regular crystal lattice, the atoms bond in irregular and nonrepeating patterns. These materials are referred to as amorphous solids.

As a result of this unusual bonding, amorphous solids have a very wide range of properties. Some amorphous solids, such as window glass, are hard, brittle, and have a high melting point, while other amorphous solids, such as rubber or plastic, are soft and have very low melting points.

The Least You Need to Know

- The atoms or molecules in solids are locked into place by a variety of different methods.
- Crystals are the repeating patterns of atoms or molecules in solids. There are many types of crystal structure, but one thing they all have in common is that the patterns are repeated over and over again to give the crystal long-range order."
- The properties of solids depend largely on the method used to hold the atoms or molecules in place.

Chapter 13

Liquids and Intermolecular Forces

In This Chapter

- Properties of liquids
- The three types of intermolecular forces
- How intermolecular forces play a role in the properties of liquids

In the last chapter, we discussed the properties and types of solids. It seems only fitting that this chapter is about liquids. For those of you who are good at spotting patterns, yes, we'll soon be learning about the properties of gases in Chapter 15.

If you've ever taken a chemistry class, you've probably noticed that many chemical reactions take place in the liquid phase. The reason for this is simple: liquids are easy to work with. Four out of five chemists prefer to work with chemicals in some liquid form because they're easy to measure, they don't require much special equipment to handle them, and they mix with each other fairly quickly. Liquids are the perfect medium for chemistry!

Of course, because liquids are used for a lot of things, it's important that you know how they behave. That's where this chapter comes in!

What's a Liquid?

This seems like a silly question because we all have a good feeling for what a liquid is. If I throw something on you and it pokes you in the eye, it's probably not a liquid. On the other hand, if it makes your shirt feel wet and sticky, it probably is.

Liquids are the phase of matter in which molecules can move around freely, but still experience forces that keep them together. For example, if you pour a glass of water on the floor, the water will tend to pool rather than spread into an infinitely thin layer. As a result, we classify water as a liquid under standard conditions.

def•i•ni•tion

A **liquid** is the form of matter in which molecules move around freely but still experience attractive forces.

A general property of liquids is that their volumes remain constant even if their shapes change. For example, you can easily put 500 mL of water into a rubber glove; when you stretch the rubber glove, the shape of the water changes but the volume doesn't.

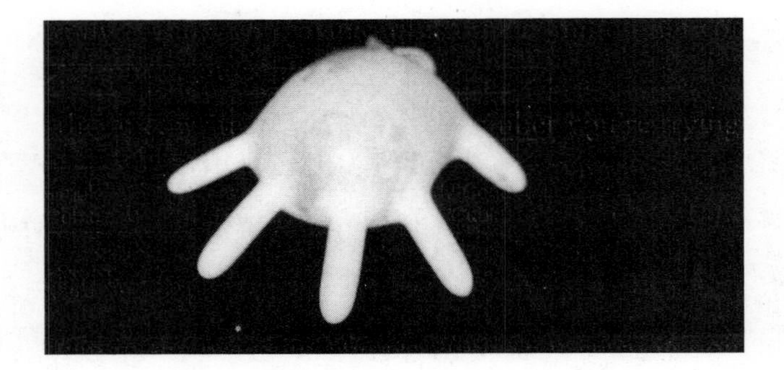

Figure 13.1

This rubber glove contains 500 mL of water. No matter what shape it's forced into, the volume of the water remains the same.

Intermolecular Forces

You may be familiar with the "ball pits" that can frequently be found in the play areas at fast food restaurants to keep little kids amused. They contain thousands of hollow plastic balls in a big pit that kids can jump around in and throw at each other. Well, liquids are like these "ball pits." With very little force, the balls can be moved from one place to another because they don't stick very tightly to surrounding balls.

Now, you may have noticed if you've ever visited one of these ball pits that the balls *do* stick together. Little kids, after all, aren't terribly careful to make sure that they don't jump into these pits covered with pizza grease and soda, so over time the balls tend to stick to one another. Similarly, the molecules in liquids are also a little bit "sticky" and tend to hang around one another. However, this isn't due to soda pop and runny noses—it's due to something called "intermolecular forces." Let's take a look at how these intermolecular forces work.

Dipole-Dipole Forces

Many covalent molecules stick together like little magnets. One side of the molecule has some positive charge on it while another side of the molecule contains some negative charge. Generally, things that have both positive and negative charge on them are referred to as being "polar."

Consider NF_3, for example. Using the rule that electronegativity increases as you move from left to right across a row of the periodic table (Chapter 7), fluorine is more electronegative than nitrogen, meaning that it pulls on electrons more strongly than nitrogen. Let's examine what this looks like in a single N-F bond:

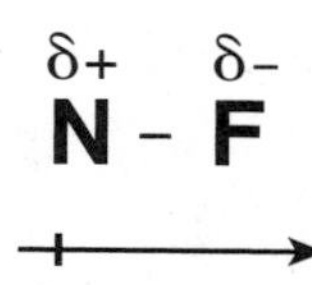

Figure 13.2

Because fluorine is more electonegative than nitrogen, it pulls some of the electrons away from it and has a partial negative charge.

Let's learn what the symbols in this figure mean. Fluorine is more electronegative than nitrogen, so it pulls harder on the electrons in the covalent bond than nitrogen does. As a result, the electrons tend to spend more time hanging around the fluorine atom than the nitrogen atom. This uneven sharing of electrons causes fluorine to have a partial negative charge, denoted by the symbol $\delta-$.

Likewise, if fluorine is pulling electrons away from nitrogen, then nitrogen has fewer electrons hanging around it, giving it a partial positive charge, δ^+. Because the electrons in this bond are distributed unevenly, it's referred to as being a "polar covalent bond." Polar covalent bonds are formed whenever two elements with dissimilar electronegativities form covalent bonds.

The Mole Says

A good way to remember that the dipole arrow points toward negative instead of positive is that the back end looks like an addition sign. This addition sign always points toward the partial positive charge in the molecule, while the head of the arrow points toward the partial negative charge.

Let's examine the structure of the entire NF_3 molecule.

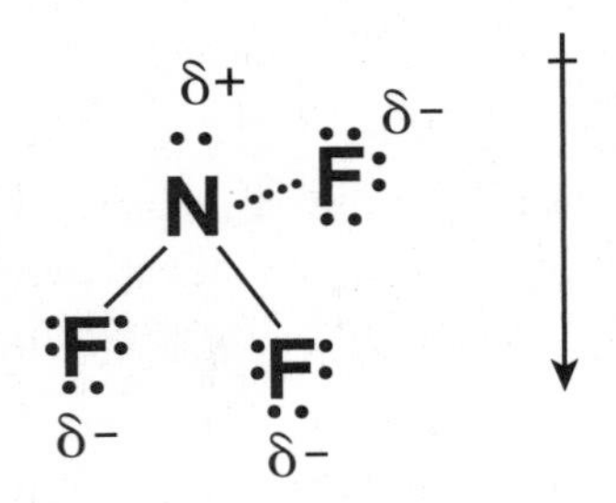

Figure 13.3

Each fluorine atom pulls electrons away from nitrogen. As a result, the side of the molecule containing the three fluorine atoms gains a partial negative charge, making the whole molecule polar.

As you can see, the three polar N-F bonds are all pointing in the same direction. Because the side of the molecule containing the fluorine atoms has three partial negative charges and the side of the molecule containing the nitrogen atom has a partial positive charge, the whole molecule is polar.

Similarly, we can see that OF_2 is a polar molecule because both fluorine atoms contain a partial negative charge:

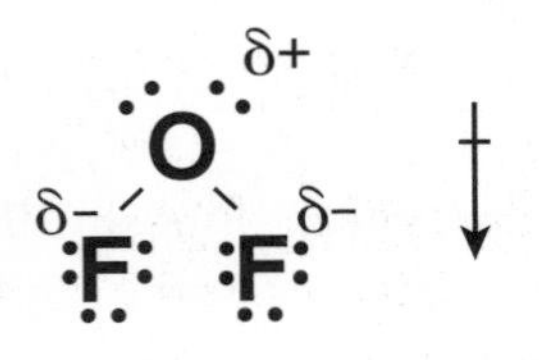

Figure 13.4

Both fluorine atoms pull electrons from oxygen, causing both of the O-F bonds to be polar and the whole OF_2 molecule to be polar.

Bad Reactions

Not all molecules that contain polar bonds are necessarily polar molecules. For example, if all the bonds are arranged at equal angles from the center of the atom (as in a Mercedes hood ornament or the points of a pyramid), the charges will be pulled equally in all directions, canceling each other out.

You've Got Problems

Problem 1: Draw the Lewis structure of phosphorus trichloride and determine the partial charges on each atom and the overall dipole for the molecule.

In a liquid containing polar molecules, the side of the molecule with partial positive charge will align itself with the partially negative side of neighboring molecules. The attractive force between the molecules that results from this interaction is referred to as a *dipole-dipole force*. An example of how this looks is shown in the following figure.

Dipole-dipole forces, while strong enough to keep the molecules in a liquid together, are much weaker than ionic or covalent bonds. As a result, covalent molecules are able to move freely throughout the liquid.

def•i•ni•tion

Dipole-dipole forces are the attractive forces that hold polar molecules together. This attraction is caused when the partially negative side of one molecule interacts with the partially positive side of another.

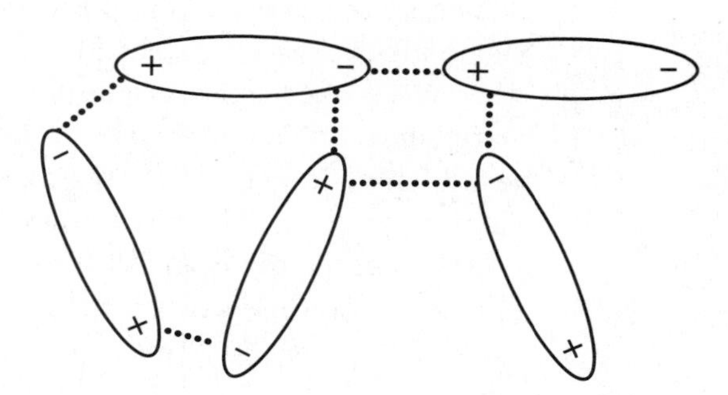

Figure 13.5

Polar molecules align themselves to maximize the number of attractions between opposite charges and minimize the number of repulsions between similar charges.

Hydrogen Bonding

Some very polar covalent compounds contain a hydrogen atom bonded to a nitrogen, fluorine, or oxygen atom. As a result, the hydrogen atom on one molecule (which has a high partial positive charge) has a very strong attraction to the lone pair electrons on the N, F, or O atom on a neighboring molecule. This very strong force is referred to as a *hydrogen bond.*

To understand why this happens, let's use the example of hydrofluoric acid, HF. In HF, we have a very polar H-F bond due to fluorine's extremely high electronegativity. As a result, most of the electrons are pulled toward fluorine, leaving very little electron density around hydrogen.

def•i•ni•tion

The term **hydrogen bond** isn't technically accurate. Though the interaction between hydrogen and highly electronegative atoms is strong for an intermolecular force, it's nowhere near as strong as a true chemical bond. However, since everybody uses this term, it should be used to describe these interactions.

$$\underset{\delta+}{H}-\underset{\delta-}{\ddot{\underset{\cdot\cdot}{F}}}: \dashrightarrow \underset{\delta+}{H}-\underset{\delta-}{\ddot{\underset{\cdot\cdot}{F}}}:$$

Figure 13.6

In HF, fluorine pulls most of the electron density from hydrogen. Because hydrogen has no inner electrons, the partial positive charge on it is very strong, leading to very strong interactions with the lone pairs on the fluorine atoms from other HF molecules.

> **Chemistrivia**
>
> The two strands of a DNA molecule are stuck to each other by hydrogen bonds between each base pair. Hydrogen bonds are much weaker than either ionic or covalent bonds, allowing the two strands in DNA to "unzip" from each other with relative ease.

Because hydrogen has very little electron density, it has a partial positive charge. However, unlike other elements that have a partial positive charge, hydrogen has no inner electrons to shield the nucleus from other atoms. As a result, atoms with partially negative charge have extremely strong electrostatic interactions with hydrogen. These hydrogen bonds, while still not as strong as covalent bonds or the attractive forces between anions and cations, are much stronger than other intermolecular forces.

London Dispersion Forces

So far, we've seen the forces that bind together polar molecules in a liquid. What kind of forces hold the molecules in a nonpolar liquid together? You may be surprised to find that nonpolar molecules also depend on the attraction of opposite charges to stay together in a liquid.

How does this process work? After all, nonpolar molecules, by definition, don't normally have any partial positive or negative charge under normal circumstances! The following figure shows how this works when helium is liquefied:

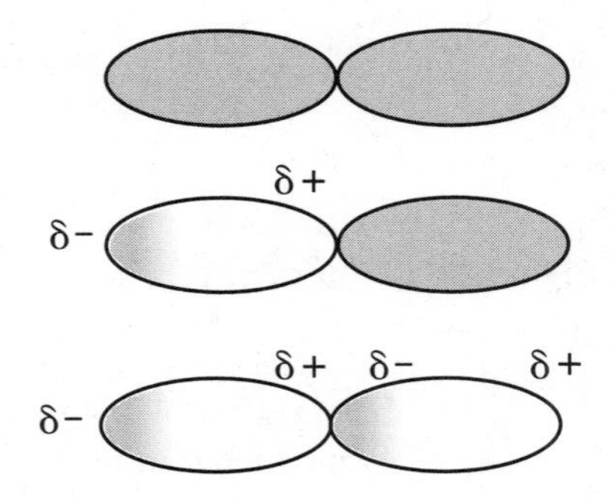

Figure 13.7

London dispersion forces are created when one molecule with a temporary dipole causes another to become temporarily polar.

In the top illustration, we see two helium atoms next to each other. As expected, neither of the atoms has a partial charge. However, if the electrons on one of the helium atoms were to temporarily move to one side of the helium atom, this atom would become very temporarily polar—this is shown in the second illustration. Because the atom on the left is polar, the electrons on the helium atom on the right are attracted toward it, causing the second atom to also become temporarily polar. The attractive force of these two temporary dipoles is referred to as a London dispersion force.

Similarly, these temporarily induced dipoles can take place in nonpolar molecules as well. For example, two methane molecules can undergo the same process and become attracted to each other by London dispersion forces.

As you might expect, this effect is very temporary, because the random movements of electrons within an orbital quickly cause the temporary dipole to disappear. As a result, this very weak, very short-lived force is nowhere near as strong an interaction as either dipole-dipole forces or hydrogen bonds.

Chemistrivia

London dispersion forces are strongest between very large molecules because the area of the molecule that can become temporarily polarized is larger.

The Mole Says

To recap: Hydrogen bonds are the strongest intermolecular force, dipole-dipole forces are of intermediate strength, and London dispersion forces are weakest. None of these three forces is anywhere near as strong as covalent bonds or the attractions between cations and anions.

You've Got Problems

Problem 2: Determine the intermolecular force that's strongest in each of the following compounds:

(a) PBr_3

(b) CO_2

(c) NH_3

The Effects of Intermolecular Forces

The intermolecular forces present in a compound play a role in that compound's properties. This isn't really surprising when you think about it. After all, if the molecules in one liquid are held tightly together by a strong intermolecular force, this liquid would be expected to behave differently than a second liquid in which the molecules are held together very weakly. The following are two of the ways in which intermolecular forces affect the properties of a liquid:

- Melting and boiling point: Generally, compounds that undergo hydrogen bonding melt and boil at higher temperatures than compounds that experience dipole-dipole forces or London dispersion forces. For example, let's consider the following three molecules.

Compound	Intermolecular Force	Melting Point	Boiling Point
CH_4	London dispersion force	–182° C	–164° C
HCl	Dipole-dipole	–115° C	–85° C
H_2O	Hydrogen bonding	0° C	100° C

- Surface tension: the tendency of liquids to keep a low surface area. Liquids with stronger intermolecular forces tend to have higher surface tension than those with weak intermolecular forces. For example, if you pour a very small amount of water on a table, it will tend to collect together in one large drop. On the other hand, if you do the same thing with gasoline, which has weaker intermolecular forces, you'd find that the gasoline tends to spread out over a larger area. For those of you who were planning on pouring gasoline on your coffee tables, please only do so under the supervision of a grown-up!

You've Got Problems

Problem 3: Rank the following compounds from lowest to highest boiling point, based on the intermolecular forces involved: PF_3, HF, CF_4.

The Least You Need to Know

- Liquids have a fixed volume but no fixed shape.
- The atoms or molecules in a liquid can easily move around one another, but are held together by weak attractive forces called intermolecular forces.
- Dipole-dipole forces are attractions between the partial opposite charges on two polar molecules.
- Hydrogen bonds are unusually strong dipole-dipole attractions that occur when hydrogen bonds to nitrogen, oxygen, or fluorine.
- London dispersion forces are attractions between the temporarily induced dipoles on two nonpolar molecules.
- The strength of the intermolecular force in a material affects its melting point, boiling point, and surface tension.

Chapter 14

Solutions

In This Chapter

- What solutions are
- Why some things dissolve and some things don't
- Concentration
- Factors affecting solubility
- Dilutions

As I mentioned in the beginning of Chapter 13, chemists prefer to work with liquids because they're easy to handle. However, if you take a look at the chemicals in your cupboard (baking soda, sodium chloride, Twinkies, etc.), you find that many of them are solids. Though solids are also easy to measure and store, they don't react anywhere nearly as quickly as liquids do. By dissolving solids in liquids, one can manipulate their concentrations to ensure quick reaction rates. The resulting mixtures are referred to as solutions.

However, not all liquids dissolve all solids. If they did, your drinking glass would dissolve every time you poured yourself a refreshing glass of milk. In this chapter, we'll learn about the formation, behavior, care, and feeding of solutions.

What Are Solutions?

The word "solution" is just another fancy term for homogeneous mixture (see Chapter 6). In solutions, one material (called the solute) is completely dissolved in another (called the solvent). Examples of solutions that I use around my house every day are fruit punch and contact lens solution, both of which contain solid solutes dissolved in water.

Chemistrivia

When two liquids mix to form a solution, we usually define the minor component as being the solute and the major component as the solvent. In cases where two liquids are completely soluble in one another at all concentrations, we say that they are "miscible" (pronounced "miss-eh-ble"). For example, rubbing alcohol and water are miscible with one another. Though there are many thousands of different liquids, you can't place more than four different liquids in a container before you find two that start to mix.

How and Why Do Things Dissolve?

When making a solution, it's handy to know if one thing will dissolve in another. After all, if somebody wants you to make them a liquid solution of one chemical, it won't make you look good if you bring them a beaker of liquid with sludge sitting at the bottom because you chose the wrong solvent.

The Mole Says

The phenomenon that polar solvents dissolve ionic and polar solutes, nonpolar solvents dissolve nonpolar solutes, and polar solvents don't dissolve nonpolar solutes (and vice-versa) is often summed up by the phrase "Like dissolves like."

The best way to tell if something will dissolve is to look at the polarities of the solvent and the solute. If the polarities of the solvent and solute match (both are polar or both are nonpolar), then the solute will probably dissolve. If the polarities of the solvent and solute are different (one is polar, one is nonpolar), the solute probably won't dissolve. Let's explore why this happens.

Why Polar Solvents Dissolve Ionic and Polar Solutes

As mentioned earlier, polar solvents are good at dissolving polar solutes. To explain this, we'll describe the process that occurs when table salt (sodium chloride) dissolves in water.

As we learned in Chapter 11, water is a polar molecule with partial positive charge on each hydrogen atom and partial negative charge on the oxygen atom. This polarity is shown in the following figure:

Figure 14.1

Water is a polar covalent molecule that's good at dissolving polar solids.

Ionic solids like sodium chloride, by definition, contain cations and anions. As a result, when an ionic solid such as sodium chloride is placed into water, we see the following take place:

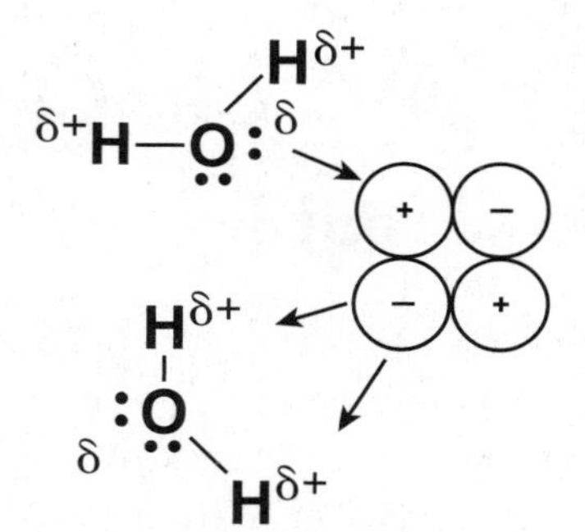

Figure 14.2

Sodium chloride is highly soluble in water.

When sodium chloride is placed into water, the partial positive charges on the hydrogen atoms in water are attracted to the negatively charged chloride ions. Likewise, the partial negative charges on the oxygen atoms in water are attracted to the positively charged sodium ions. Because the attractions of the water molecules for the sodium and chloride ions are greater than the forces holding the crystal together, the salt dissolves. When a solute dissolves in water, the process is referred to as hydration.

Chemistrivia

In some cases, the attraction of water molecules for the polar solute isn't strong enough to pull the solute molecules apart. As a result, some polar solutes don't dissolve in water.

Similarly, we find that polar solutes such as methanol, ethanol, and isopropanol are highly soluble in water because both are highly polar.

Why Polar Solvents Don't Dissolve Nonpolar Solutes

The "like dissolves like" rule indicates that polar solvents will do a poor job of dissolving nonpolar solutes. We can understand this by looking at the following figure.

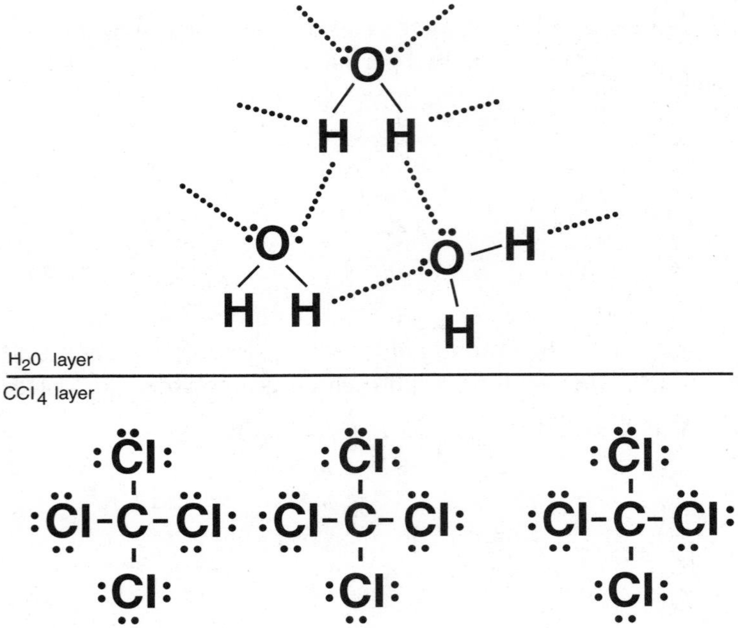

Figure 14.3

Water doesn't dissolve carbon tetrachloride because the strong interactions between water molecules are more important than the weak interactions between water and carbon tetrachloride.

H_2O layer

CCl_4 layer

In the preceding figure, we can see what happens when we place carbon tetrachloride into water. Because carbon tetrachloride is a nonpolar molecule, the interactions between adjacent molecules are very weak. As a result, we might expect carbon tetrachloride to be very soluble in water. However, water molecules form strong hydrogen bonds with one another, causing them to stick tightly to one another. Since the water molecules have very strong intermolecular forces with each other and interact only weakly with carbon tetrachloride (via London dispersion forces—see Chapter 13), CCl_4 is almost completely insoluble in water.

Why Nonpolar Solvents Don't Dissolve Polar Solutes

Let's imagine what happens when a polar solute such as sodium chloride is placed in a nonpolar solvent such as carbon tetrachloride. Because CCl_4 doesn't have a partial charge, it won't attach itself to the sodium or chloride ions. As we've mentioned before, the sodium and chloride ions in NaCl are strongly attracted to one another because of their opposite charges. This very weak solvent-solute interaction, as well as the very strong attraction between neighboring solute particles, causes sodium chloride to be insoluble in carbon tetrachloride.

Why Nonpolar Solvents Dissolve Nonpolar Solutes

If we place a nonpolar solid into a nonpolar liquid, "like dissolves like" implies that the solid will dissolve. However, the only forces that will cause the liquid to be attracted to the solid are weak London dispersion forces. Why should the solid dissolve?

Let's imagine that we have placed a chunk of carbon tetrabromide in a beaker containing carbon tetrachloride. The carbon tetrabromide molecules in the solid are held together by very weak London dispersion forces, as are the carbon tetrachloride molecules in the solvent. One might expect, then, that there is no particular reason for the solute to dissolve.

As it turns out, there's another force involved. Processes that increase the randomness of a system usually occur spontaneously (we'll discuss this phenomenon, known as entropy, in Chapter 26). Because the molecules in carbon tetrabromide will be made more random if they're mixed with another compound, the carbon tetrabromide will dissolve in the carbon tetrachloride.

You've Got Problems

Problem 1: Based on the polarity of each, determine whether the solvent in each of the pairs will dissolve the solute listed.

a) Solvent: water. Solute: lithium chloride.

b) Solvent: chloroform ($CHCl_3$). Solute: methane.

c) Solvent: chloroform ($CHCl_3$). Solute: water.

Determining the Concentration of a Solution

There are many ways to measure the amount of solute present in a solution. Each method is useful for a different purpose in chemistry, so we're unfortunately stuck with the task of learning all of them. Without further ado, here they are.

Qualitative Concentrations

The amount of solute present in a solution can be described without numbers by one of the following terms:

- **Unsaturated.** "Unsaturated" refers to any solution that is still capable of dissolving more of a solute. For example, a glass of iced tea is not saturated with sugar if you've placed one tablespoon of sugar in it because it's still capable of dissolving more sugar. This term isn't very good for determining the exact quantity of solute present—for example, both a glass of water and a filled swimming pool would be said to be unsaturated salt solutions if there were one gram of salt dissolved in each.
- **Saturated.** These solutions have dissolved the maximum possible amount of solute. For example, if you keep adding sugar to a glass of Kool-Aid, it will

eventually stop dissolving and settle to the bottom (little kids, however, refuse to believe this). This solution is said to be saturated.

- **Supersaturated.** These solutions are those that have dissolved *more* than the normal maximum possible amount of solute. These solutions are unusual and aren't very stable. For example, the addition of a small mote of dust to such a solution causes enough of a disturbance that crystals spontaneously form until the solution reaches a saturated state.

It's easy to tell if a solution is unsaturated, saturated, or supersaturated by adding a very small amount of solute. If the solution is unsaturated, the solute will dissolve. If the solution is saturated, it won't. If the solution is supersaturated, crystals will very quickly form around the solute you've added.

Molarity (M)

Molarity is probably the most commonly used way of measuring concentration and is defined as the number of moles of solute per liters of solution.

Bad Reactions

When reading the definitions of the following methods of determining concentration, pay close attention to whether the volume component asks for the weight or volume "of solution" or weight or volume "of solvent." When the weight or volume of the solution is specified, it means that you're interested in the amount of solution present *after* the solute has been added. If the weight or volume of the solvent is specified, this means that you're interested in the amount of solvent *before* the solute has been added.

You've Got Problems

Problem 2: What is the molarity of a solution with a volume of 3.0 liters that contains 120 grams of acetic acid ($C_2H_3O_2H$)?

Let's say that we have made a solution by adding water to 120 grams (3.0 mole) of sodium hydroxide until the final volume of the solution is 1.0 liters (to review mole calculations, head back to Chapter 11). Because we have 3.0 moles of solute in 1.0 liters of solution, the molarity is equal to (3.0 mole)/(1.0 liter) = 3.0 M. We refer to a solution with a molarity of 3.0 as being a "3.0 molar" solution.

Molality (m)

Molality is defined as the number of moles of solute per kilogram of solvent. For example, if we were to add two kilograms of water to 4 moles of sugar, the molality would be equal to 4 moles/2 kilograms = 2 m ("two molal"). When doing calculations with molality, note that because the density of water is 1.0 g/mL under standard conditions, the number of kilograms of water is equal to the number of liters of water.

You've Got Problems

Problem 3: Determine the molality of a solution in which 45 grams of calcium acetate are added to 560 mL of water.

Normality (N)

The normality of a solution is defined as the number of moles of a reactive species, usually referred to as "equivalents" per liter of solution. The use of "equivalents" will depend on the reaction being performed, so some knowledge of the specific chemical process in a reaction is necessary before computing normality. At least, that's the "normal" way of solving this problem. (I couldn't resist.)

Mole Fraction (χ)

The mole fraction is defined as the number of moles of one component in a solution divided by the total number of moles of all components in the mixture. In equation form, we can express the mole fraction of one component in a solution as being:

You've Got Problems

Problem 4: What is the mole fraction of water in a solution made by mixing 4.5 moles of isopropanol with 15.0 moles of water?

$$\chi_A = \frac{\text{moles of A}}{\text{moles of A + moles of B + moles of C}} + \ldots$$

where A refers to the first component, B refers to the second component, and C refers to the third component. As the "…" indicates, this calculation can be extended to include any number of components in the mixture.

Parts Per Million (ppm) and Parts Per Billion (ppb)

Both parts per million and parts per billion are units of concentration most frequently used in environmental analysis. Because the solvent used is most frequently water, the concentration of a solution in ppm can be found by dividing the number of mg (0.001 g) of solute by the number of liters of water. Parts per billion can be determined by dividing the number of μg (10^{-6} g) of solute by the number of liters of water.

A Quick Summary of Units of Concentration

The following table includes all of the units of concentration we've mentioned in this chapter, as well as how to find them.

Unit	Symbol	How It's Measured
molarity	M	moles of solute / liters of solution
molality	m	moles of solute / kilograms of solution
normality	N	"equivalents," which varies depending on the reaction being performed
mole fraction	χ	$\frac{\text{moles of A}}{\text{moles of A + moles of B + ...}}$
parts per million	ppm	mg solute/L of water
parts per billion	ppb	µg solute/L of water

Factors That Affect Solubility

Sometimes we want a thing to dissolve quickly because we get bored sitting in front of a beaker watching it. Sometimes we want a larger quantity of a solute to dissolve than we could normally achieve. Before this chapter, both of these things would be impossible. However, now that you understand how solutions work, I feel confident in handing you the following ways of affecting solubility.

Surface Area of the Solute

Let's imagine that you're trying to dissolve 1.0 grams of sodium chloride in a glass of water. Which would dissolve more quickly: a large 1.0 gram crystal or 1.0 grams of salt ground into a fine powder?

If you guessed that the powder would dissolve more quickly, you're right. Because the powder has a larger surface area than the crystal, more of the ions in the salt are exposed to the solvent at any given time, causing them to dissolve more quickly. It should be noted that breaking a solute into smaller pieces doesn't change how much of it will dissolve, it only changes how quickly it will dissolve.

Pressure

When dissolving a gas within a liquid, the pressure of the gas has a huge effect on its solubility. When the pressure of a gas is low, the number of gas molecules that hit the surface of the liquid at any given time is low; as a result, there are fewer chances for the gas to dissolve. However, if the pressure of the gas is increased, the number of collisions between the gas molecules and the solvent increases, which causes more of the gas molecules to dissolve.

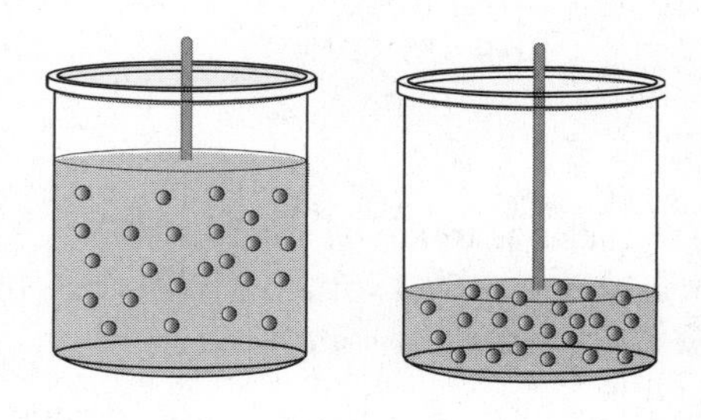

Figure 14.4

(a) In the first piston shown, the pressure of the gas is low and the gas molecules don't collide with the solvent very frequently. (b) If the pressure is increased in the piston, the gas molecules will undergo more frequent collisions, leading to higher solubility.

The relationship between pressure and solubility was first expressed in 1801 in Henry's Law:

$$P = kC$$

In this equation, P represents the pressure of the gas above the solvent, k is a mathematical constant with positive value that depends on the particular solution being studied, and C represents the concentration of the gaseous solute in the solution. As you can see from the equation, the higher the pressure of the gas, the more concentrated the solution will be.

Chemistrivia

When divers are deep underwater, the pressure of the water increases the amount of gas from the air they breathe that dissolves in their blood. As they rise, the pressure goes down and the gas becomes less soluble, forming small bubbles in the bloodstream. Because gas bubbles accumulate in one's joints, rising too quickly can increase this amount of gas to dangerous levels. The agony caused by this condition causes people to curl up in the fetal position, so it's called "the bends."

Though pressure is an important factor in the solubility of a gas, pressure has very little effect on the solubilities of liquids or solids.

Temperature

The temperature of a liquid affects the solubility of both solids and gases. Generally, increasing the temperature of a solvent increases the solubility of most ionic compounds (though there are exceptions). More important, increasing the temperature of a solvent usually increases the rate at which a solute dissolves, which is why it's easier to dissolve sugar in hot tea than cold tea. Gases, on the other hand, become less soluble as the temperature of the solution increases, which is why carbonated beverages (which contain CO_2) go flat more quickly on hot days than on cold ones.

Dilutions

Let's say that I want to make as much of a 0.100 M NaCl solution as I can, but the only thing in my stockroom is 1.50 L of a 0.450 M NaCl solution. How would I turn the 0.450 M solution into a 0.100 M solution?

I would dilute it! Dilution is the process by which a solvent is added to a solution to make the solution less concentrated. The equation we'd use for dilutions is:

$$M_1V_1 = M_2V_2$$

M_1 is the initial molarity of the solution, V_1 is the initial volume of the solution, M_2 is the molarity of the solution after it has been diluted, and V_2 is the volume after dilution.

To determine how we'd make my 0.100 M NaCl solution, we'd use the equation above, inserting the values that I know:

$$M_1V_1 = M_2V_2$$

$$(0.450\ M)(1.50\ L) = (0.100\ M)V_2$$

$$V_2 = 6.75\ L$$

I can make 6.75 L of a 0.100 M NaCl solution.

You've Got Problems

Problem 5: How much of a 0.500 M NaCl solution would be needed to make 750 mL of a 0.125 M solution?

The Least You Need to Know

- A solution is the same thing as a homogeneous mixture.
- The solute in a solution is the thing that gets dissolved, and the solvent is the thing that does the dissolving.
- "Like dissolves like."
- There are many different ways to express the concentration of a solution.
- The surface area of the solute, the pressure, and the temperature all affect solubility.
- Dilutions occur when you add extra solvent to a solution to decrease its concentration.

Chapter 15

The Kinetic Molecular Theory of Gases

In This Chapter

- Properties of gases
- The Kinetic Molecular Theory of Gases
- Ideal gases
- Root mean square velocity
- Effusion and diffusion

Gases are sometimes more difficult to visualize than other forms of matter. If you're anything like me, you learned at an early age that solids are hard when you bashed your face on the coffee table, spraying blood all over your grandparents' house while the fourteen-year-old babysitter cried. You learned that liquids are wet when you were two and your parents let you sleep through the night in your big boy pants instead of diapers, and you responded by wetting the bed. Both of these phases of matter are easy to relate to because you can interact with them in easily understood ways.

On the other hand, you may not have learned about gases until you went to school and realized that the stuff you breathe has substance, even though you can't see it, smell it, taste it, or bash your head on it. In this chapter, we're going to investigate both the basic properties of gases (with which you're probably already somewhat familiar) as well as learn why gases have these properties.

What Are Gases?

Gases are the phase of matter in which particles are usually very far apart from one another, move very quickly, and aren't particularly attracted to one another. Because the molecules in a gas are so far apart from one another, gases are much less dense than liquids or solids. For example, it's easier to pick up a huge balloon full of air than a huge water balloon.

We can see the differences between the structures of solids, liquids, and gases by looking at the following figure:

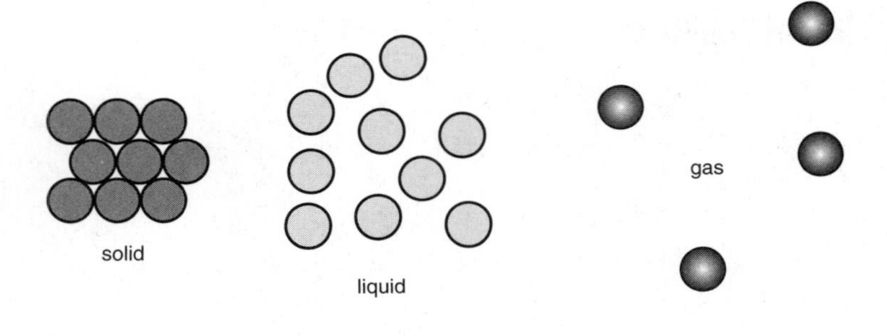

Figure 15.1

(a) Solids consist of particles that are bound in place by a variety of relatively strong forces. (b) In liquids, the particles can move freely past one another because they are only held together by weak intermolecular forces. (c) The particles in a gas have almost no attractive forces at all, which allows them to spread out.

The reason that gases ignore any intermolecular forces (Chapter 12) that might normally exist between the atoms or molecules is that they have enough energy to overcome the strength of these forces. For example, elements that experience weak intermolecular forces vaporize at extremely low temperatures (in the case of helium, –269° C). On the other hand, if a chemical compound is held together by strong intermolecular forces such as the hydrogen bonds in water, a much larger amount of energy is required to overcome these forces (H_2O boils at 100° C).

The following are general properties of gases:

- *Gases don't have a fixed shape.* Gases fill the nooks and crannies of whatever container you put them in.
- *Gases don't have a fixed volume.* Unlike liquids, gases expand across an area until something physically stops them. This phenomenon explains why little old ladies who wear too much perfume can frequently be smelled long before you see them coming.
- *Gases mix freely with other gases.* Unlike liquids, which sometimes don't mix at all (e.g., oil and water), any combination of gases will always mix with one another.
- *Gases can be easily compressed.* Because there's a lot of space between the molecules in a gas, you can easily squish them down. Solids and liquids, on the other hand, are much less compressible.

The Kinetic Molecular Theory of Gases: Why Gases Do What They Do

As mentioned before, gases are harder to visualize than other phases of matter. This is true not only because it's difficult to see and study them, but because all the molecules in a gas behave independently of one another. As a result, studying the behavior of a gas containing one mole of molecules that are flying all over the place is a lot harder than studying the behavior of a crystal containing a mole of stationary molecules.

Because it's tough to study all of the particles in a gas, scientists have come up with a variety of theories to simplify gases' behavior so they can be more easily understood. Probably the most important of these theories is referred to as the Kinetic Molecular Theory (KMT).

Bad Reactions

All theoretical models (including the KMT) only approximate the behavior of the thing being modeled. The approximations that define each model are designed to make the real phenomenon easier to understand and predict. However, no model is perfect, which explains why weather forecasting models usually get the five-day forecast wrong.

The KMT makes the following assumptions about the behavior of the particles in a gas. Again, these assumptions are not always true, but allow us to understand gases more easily.

The Particles in a Gas Are Infinitely Small

We've mentioned on a number of occasions that atoms and molecules are really, *really* small (~10^{-10} m, a unit referred to as an "angstrom"). The Kinetic Molecular Theory not only states that atoms and molecules are small, but says they have no volume at all!

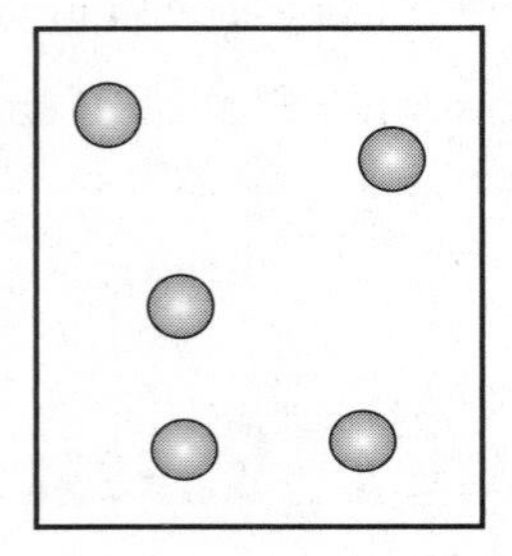

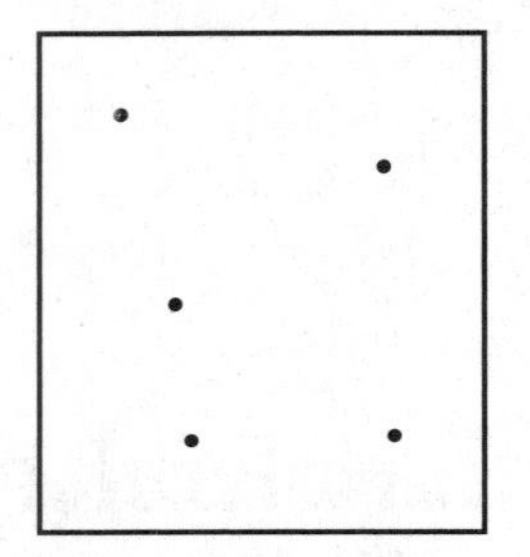

Figure 15.2

The molecules in real gases look like the box on the left. The KMT assumes that molecules have no volume at all, like the box on the right. Is this perfectly accurate? Nope. Is this almost accurate? Yep.

Chemistrivia

Gases are compressible because gas molecules are so small compared to the volume of the gas. If the molecules took up more space, they couldn't squish together as easily.

We make this assumption because it makes the mathematics behind the model easier to follow and because gas molecules are *really* small. For example, the water molecules in water vapor at 100° C make up approximately 1/1700 of the total volume of the gas at standard pressure. Improving the model to account for this very tiny volume wouldn't improve the accuracy of its predictions, but would make the math a lot more difficult. As a result, we just assume that the molecules are infinitely tiny.

The Particles in a Gas Are in Constant Random Motion

The KMT assumes that the particles in a gas, like very small children, constantly move from place to place in an unpredictable fashion. This assumption is correct. Furthermore, the KMT assumes that these particles travel in straight lines until they bash into something, at which point they turn around and go somewhere else. As is also the case with very small children, this assumption is true.

Chemistrivia

The assumption that gas molecules are in constant random motion explains why they have no fixed shape or volume.

Gases Don't Experience Intermolecular Forces

As we mentioned before, gas molecules fly around at very high speeds. Because intermolecular forces are relatively weak, it's rare for molecules traveling past each other at high speeds to interact strongly.

This assumption is similar to my relationship that I had with my downstairs neighbor in college. If I was sitting out on the deck and my neighbor wandered out to his car, his "magnetic personality" (as he referred to it) caused him to stop and speak to me about lobster hunting for hours on end. However, if I was riding my bike in town and I saw him coming the other way, I'd pedal faster and say "Hi-Mikehowsitgoingnicedayseeyouathome-bye!" as I zoomed by. In such instances, my extreme speed kept our interaction to a minimum.

Chemistrivia

Because gases don't experience strong intermolecular forces, all gases are able to mix freely with one another. If intermolecular forces played an important role in gas behavior, polar and nonpolar gases wouldn't be able to mix for the same reason that polar and nonpolar liquids can't mix (see Chapter 14).

The Kinetic Energies of Gas Molecules Are Directly Proportional to Their Temperatures in Kelvin

The definition of *kinetic energy* might sound complicated, but what it really means is that when you heat a gas, the molecules move more quickly. Temperature, as it turns out, is a measurement of how much motion the particles in a material have, so it makes sense that increasing the temperature will increase the motion.

def•i•ni•tion

Kinetic energy refers to energy caused by the motion of an object. The faster an object moves, the more kinetic energy it has.

The Kelvin scale of temperature is the same as the Celsius scale we've used so far throughout the book, but it's higher by 273.15° (which we'll abbreviate as 273 for simplicity's sake). We can easily convert degrees Celsius to Kelvin using the following conversion.

$$K = °C + 273$$

For example, if the temperature outside is 20° C, we can convert this to 293 K. Note that the temperature is properly written simply as "Kelvin" and not "degrees Kelvin."

Chemistrivia

The reason we use Kelvin when working with gases instead of degrees Celsius is that gases often exist at temperatures below 0° C. As a result, if we said that the kinetic energy of a gas was proportional to the temperature in degrees Celsius, the kinetic energy of any gas cooled below the freezing point of water would be negative.

Gas Molecules Undergo Perfectly Elastic Collisions

Elastic collisions are collisions in which kinetic energy is transferred from one thing to another without a loss. If you've ever played pool, you know that the balls slow down just a little bit when they hit the bumpers on the sides of the table. If these collisions were elastic, the balls would bounce off the bumpers moving exactly as quickly as they hit them, never stopping until they either hit a pocket or you got tired of watching them bounce around the table.

Why This Is Important: Ideal Gases

From the very definition of a model, we know that the Kinetic Molecular Theory of gases isn't true. Instead of telling us how gases actually behave in the real world, it gives us an idealized version of how gases should behave under perfect conditions. As it turns out, gases that follow all of the assumptions of the KMT are referred to as "ideal gases."

For those of you who haven't picked up on the idea, here's a clarification: *There is no such thing as an ideal gas! Ideal gases are imaginary! There are as many ideal gases in the universe as there are tooth fairies! Don't tell your friends, relatives, or anybody else that ideal gases are real, because they'll lock you up for being a deluded maniac!*

The concept of an ideal gas, however, is a useful one. It's very difficult to come up with rules for describing the behaviors of real gases because they come in a variety of different shapes and sizes, as well as experience different intermolecular forces to various degrees. Since most gases behave more or less like an ideal gas, we can pretend that real gases are the same as ideal gases.

You've Got Problems

Problem 1: Give three examples of an ideal gas.

Important Terms and Units

Because we've just started dealing with gases, it's time to learn some new terms for working with them. These terms will be unbelievably handy for the rest of this book, so make sure that you understand them before moving on!

Pressure

Pressure is defined as being the amount of force exerted by the particles in a gas as they hit the sides of a container. Let's imagine that we have five gas particles in a box, as shown in the following figure.

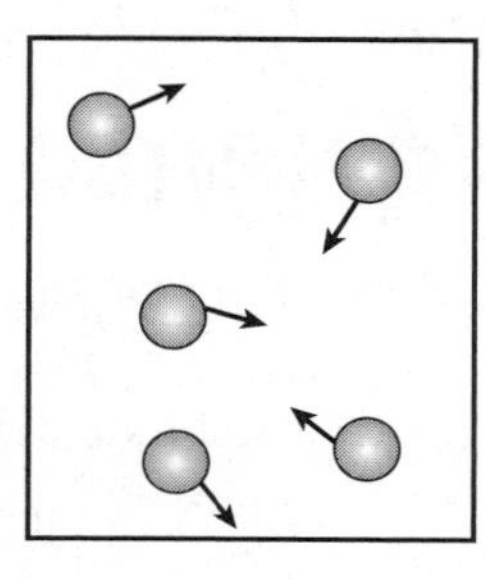

Figure 15.3

Gas pressure is caused by the force of gas molecules crashing into the sides of the container that holds them.

As the kinetic molecular theory described, these particles go whizzing around at unbelievably high speeds in random directions. Whenever the particles hit the walls, the force of the impact results in gas pressure. For those of you who want to experience this phenomenon at home, have a friend throw tennis balls at you for a few minutes. The force that drives you backward is the pressure of the tennis balls.

There are several different units of pressure that are commonly used.

- **Atmospheres (atm).** 1 atm is defined as the average atmospheric pressure at sea level. Though it's not an official metric unit, it's frequently used because it's so darn handy.
- **Millimeters of mercury (mm Hg) or torr.** Both of these units are identical. There are 760 mm Hg or torr in 1 atm.
- **Pascals (Pa).** This is the metric unit of pressure. There are 1.01325×10^5 Pa in 1 atm. Typically, we use the unit "kilopascals" (kPa) instead of pascals. There are 101.325 kPa in 1 atm.

Chemistrivia

The unit "torr" was named after Evangelista Torricelli, an Italian physicist who worked with gases. Among other things, Torricelli is credited with inventing the barometer.

- **Bar.** The bar is a unit of pressure most commonly used by meteorologists. There are 1.01325 bars in 1 atm.
- **Pounds per square inch (psi).** Though not a metric unit, it's still in widespread use for many purposes including automobile tire pressures. Unlike other units, psi indicates how much gas pressure is present in excess of one atm. There are approximately 14.7 psi in 1 atm.

Volume and Temperature

When working with gases, volume is expressed in "liters" (L). Temperature is expressed in Kelvin (K).

Other Miscellaneous Terms

In addition to the preceding terms, there are several other terms that will come in handy when working with gases.

- Standard Temperature and Pressure (STP): STP is the most common reference condition for expressing the properties of gases. Standard temperature is defined as 0° C (273 K), and standard pressure is 1 atm.
- The ideal gas constant (a.k.a. "universal gas constant"): Shown as "R" in equations. It has two values that are useful for our calculations: 0.08206 L atm/mol K, and 8.314 L kPa/mol K. The value you should use for a particular problem depends on the units of pressure (kPa or atm) you're given.

How Fast Do Gas Molecules Move?

How fast do the molecules in a gas move? We've said several times that they go "really fast," but how fast is "really fast"?

To answer this question, we first need to look at some of the factors that determine the speed of gas molecules.

The Mass of the Molecules

The KMT says that the kinetic energy of the molecules in a gas is proportional only to the temperature in Kelvin. As a result, heavy objects and light ones have the same kinetic energies at the same temperature.

Let's imagine that I'm a very bad driver (this isn't much of a stretch). During the last ice storm, the roads got very slick and I drove my car into the fence surrounding my workplace at approximately five miles an hour.

When I destroyed the fence, my car was like a heavy gas molecule. Here's an interesting question: How fast would a bicyclist need to be going to destroy the fence with the same amount of kinetic energy? If you guessed "really, really fast," you're right! Because bicycles are much lighter than my car, they need a lot more speed to get the same amount of kinetic energy as my car.

Likewise, if two molecules have the same amount of kinetic energy, the lighter one will move more quickly than the heavy one. In other words, the velocity of the molecules in a gas depends on their masses!

The Temperature of the Gas

The temperature of a gas is also important in determining its molecular speed. Because the KMT states that the amount of kinetic energy is dependent on the temperature, the temperature will determine how fast the molecules go in the first place. Of course, at a given temperature lighter molecules always move more quickly than heavier ones (as we saw just a few paragraphs ago), but *all* molecules will move more quickly if we boost the temperature of the gas.

Putting It All Together: The Root Mean Square (rms) Velocity

Taking the mass of the molecules and the temperature into consideration, we find that the average velocity of the molecules in a gas can be described by a term called the "root mean square (rms) velocity." The rms velocity of a gas is calculated using the following equation.

$$u_{rms} = \sqrt{\frac{3RT}{M}}$$

In this equation, R represents the ideal gas constant (which for this equation is always 8.314 J/mol K), T represents the temperature of the gas in Kelvin, and M represents the molar mass of the compound in kilograms (see Chapter 11 for more on molar mass).

The Mole Says

You may have noticed that the units for R are different here than when we mentioned it earlier. The units "J/mol K" are equivalent to "L atm/mol K," and make the math work out better.

For example, the rms velocity for ammonia at room temperature is found by plugging the temperature (298 K) and the molar mass (0.0170 kg/mol) into this equation with the ideal gas constant. Using this equation, the rms velocity of ammonia would be:

$$u_{rms} = \sqrt{\frac{3(8.314J/molK)(298K)}{0.0170kg/mol}} = 661m/\sec$$

This is pretty darn fast!

You've Got Problems

Problem 2: What is the average velocity of hydrogen molecules at STP?

The Random Walk

Our earlier calculation found that ammonia molecules move 661 m/sec at room temperature. If ammonia moves this quickly, why don't we immediately smell it whenever our neighbor across the street mops his floor?

If ammonia molecules traveled straight from your neighbor's floor to your nose, you would smell it almost immediately. However, molecules don't travel in straight paths—rather, they bump into each other in random fashion.

To illustrate what I mean, imagine that the Olympic committee has decided to make the marathon more interesting by blindfolding all the runners. These runners will eventually finish the marathon even if they're blindfolded. Unfortunately, it will probably take them days to finish the race because they won't be traveling on a straight path—instead, they'll be bumping into trees, spectators, each other, etc. The path that these runners take is referred to as a "random walk" because they'll be traveling very quickly in random directions.

Molecules do the same thing. Though ammonia molecules travel 661 m/sec at room temperature, they take a long time to cross a room because they keep bumping into things. The length of time it takes for molecules to travel from one place to another depends not only on their rms velocity, but also on the distance between collisions, called the "mean free path."

Effusion and Diffusion

Other important behaviors of gases explained by the Kinetic Molecular Theory are effusion and diffusion. Effusion is the rate at which a gas escapes through a small hole in a container. Diffusion is the rate at which a gas travels across a room. Both of these phenomena are illustrated by the following figure.

Figure 15.4

(a) Effusion is when a gas escapes through a small hole in a container. (b) Diffusion is the rate at which a gas travels across the room.

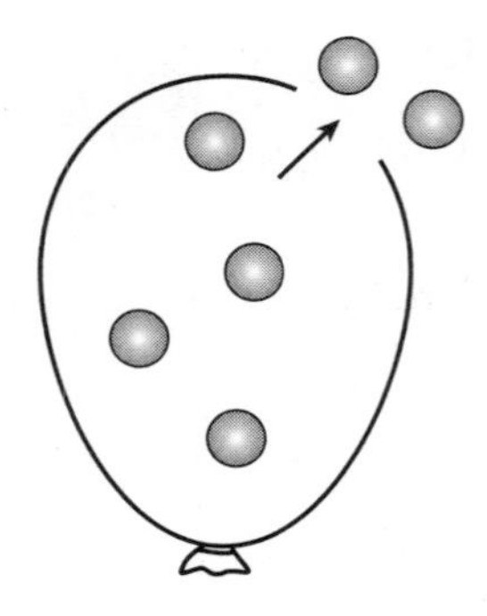

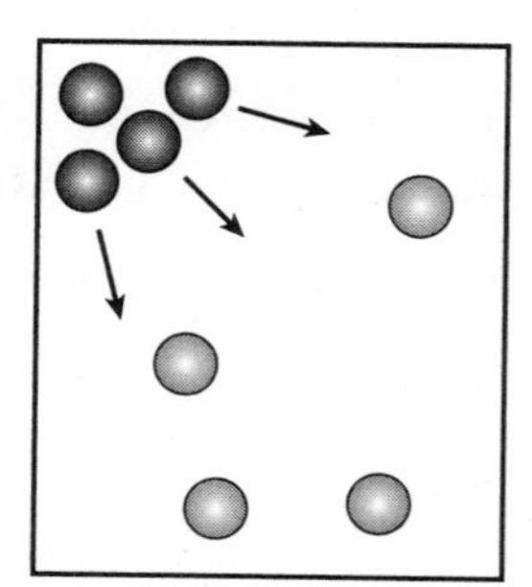

The relative rates at which two gases effuse or diffuse is explained by Graham's Law, shown here.

$$\frac{r_1}{r_2} = \sqrt{\frac{M_2}{M_1}}$$

The rate of motion of gas 1 is r_1, and M_1 denotes its molecular weight (in kg). Likewise, r_2 and M_2 stand for the rate of motion and molecular weight of gas 2. Let's see an example of how this equation is used:

Example: If I pop a balloon of methane and a balloon of carbon dioxide at the same time, which gas will diffuse to the other end of the room more quickly?

Solution: We'll let r_1 represent the rate of diffusion of methane and r_2 represent the rate of diffusion of carbon dioxide. Plugging the molecular weights of both gases into this equation, we find that:

$$\frac{r_{CH_4}}{r_{CO_2}} = \sqrt{\frac{0.0440kg/mol}{0.0180kg/mol}} = 1.56$$

We have determined that methane will travel across the room 1.56 times faster than carbon dioxide, making it the first to hit the opposite wall.

You've Got Problems

Problem 3: I filled a balloon with H_2 and found that it went flat in 16.0 hours. Using this information, how long will it take a balloon full of helium to go flat?

The Least You Need to Know

- Gases have no fixed volume or shape, are compressible, and mix freely with other gases.
- The Kinetic Molecular Theory (KMT) is a series of assumptions that's often used to approximate the behavior of real gases.
- Ideal gases don't exist, but are useful in modeling how real gases behave.
- The root mean square velocity of the molecules in a gas depends on both its temperature and molecular mass.
- Effusion and diffusion are both related to the molecular mass of a gas.

Chapter 16

Gas Laws

In This Chapter

- Boyle's Law, Charles's Law, Gay-Lussac's Law, and the combined gas law
- Avogadro's Law
- The ideal gas law
- Dalton's law of partial pressures

In the previous chapter, we spent a great deal of time discussing how gases behave on a molecular level with the Kinetic Molecular Theory (KMT). Once we developed the KMT, we were able to explain the easily observed properties of gases in terms of this theory.

Unfortunately, the calculations we did in the last chapter don't help us with most of the common problems we need to solve. For example, what happens to the pressure of a gas in a closed container when we raise the temperature from 25° C to 500° C? At first glance, this may not seem like a very interesting problem to solve. However, if you throw a can of spray paint into a campfire, you'll see a spectacular demonstration of why this is interesting. (By the way, it's a bad idea to actually attempt this demonstration—if you want to see what happens, I'm sure some bonehead has put a video clip of it on the Internet.)

Of course, scientists in the field spend relatively little time throwing compressed gases into campfires, so we'll not only learn about how gases behave, but we'll also examine some examples in which sane chemists might use these behaviors to make the world a better place.

Boyle's Law: Why Compressed Gas Is Small

Let's do a demonstration: Inflate a balloon and put it on your chair. Now, flop down on the chair as hard as you can, squishing the balloon.

When you did this demonstration, you probably found that the balloon popped when you sat on it. This wasn't really a surprise because it seems obvious that if you sit on a balloon, it will pop. My question for you: *Why* did the balloon pop?

I don't know if it's true, but I like to think that the British chemist Robert Boyle asked himself the same question back in 1662 after sitting on a balloon. In any case, he came up with a way of explaining why balloons pop when you sit on them.

For the sake of argument, let's say that your balloon had an initial volume of one liter, and that the pressure inside the balloon was initially one atmosphere. When you sat on the balloon, let's say that the force of your behind squished the volume of the balloon to 0.500 liters. What was the pressure inside the balloon after it had been squished?

Boyle's Law allows us to solve this problem.

$$P_1V_1 = P_2V_2$$

where P_1 is the initial pressure of the gas, V_1 is the initial volume of the gas, P_2 is the final pressure of the gas, and V_2 is the final volume of the gas. When using this equation, we assume that the temperature and number of moles of gas stay the same. Using the values from our example, we find that the pressure inside the balloon after we sat on it was.

$$(1.00 \text{ atm})(1.00 \text{ L}) = (x \text{ atm})(0.500 \text{ L})$$

$$x = 2.00 \text{ atm.}$$

You've Got Problems

Problem 1: If I compress 1,500 L of nitrogen at an initial pressure of 1.00 atm until the pressure reaches 450 atm, what will the new volume of this gas be?

The implication of this finding is clear. When you sat on the balloon, the pressure inside the balloon rose to 2.00 atm. Because the thin rubber of a balloon isn't strong enough to hold pressurized gas, it pops! The mystery of the popping balloon is now solved!

Chemistrivia

If you live in a highly elevated area (such as Denver), you may have noticed the effects of Boyle's Law at the supermarket. Potato chips are frequently packaged near sea level, which has a higher air pressure than exists at high altitudes. As the atmospheric pressure outside the bag decreases, the volume of the gas inside the bag increases. As a result, prepackaged snack food bags are often highly inflated at high altitudes.

Charles's Law: The Incredible Imploding Can

Let's do another demonstration. You'll need a brand new, never used, metal can with a screw-on cap. Remove the cap, place the can on the stove, and turn it to "high." After the can has been heated for about two minutes, take it off the stove with metal salad tongs and tightly screw on the cap.

Because I know that none of you actually did the demonstration (shame on you!), I'll just tell you what you would have seen—over a period of two or three minutes, the can would shrink until the sides caved in.

Way back in 1787, the French scientist Jacques Charles did exactly the same experiment while sitting around the house on a rainy day. (Editor's note: Historians believe that only the year in the preceding statement is correct.) When he observed the can imploding, he devised the following law to explain his findings.

$$\frac{V_1}{T_1} = \frac{V_2}{T_2}$$

V_1 is the initial volume of the can, T_1 is the initial temperature of the air in Kelvin, V_2 is the final volume of the can, and T_2 is the final temperature of the air (in Kelvin). We will assume that the pressure and number of moles of the air are constant. If the can has an initial volume of 5.00 liters, the temperature of the air before you took the can off of the stove was 250° C (523 K), and the temperature of the air after the can cooled was 25° C (298 K), we can use this equation to find the final volume of the can.

Bad Reactions

When working with gases, remember to always convert temperatures from degrees Celsius to Kelvin (K = °C + 273). If you don't, your answer will be wrong!

$$\frac{5L}{523K} = \frac{V_2}{298K}$$

$V_2 = 2.85$ L

What this equation means is that when the air inside the can cooled, the volume decreased, causing the can to implode. The kinetic molecular theory would explain this by saying that the air molecules had less kinetic energy at the lowered temperature, causing them to strike the sides of the can with less energy than they did before. Because the energy of the molecules hitting the sides of the can decreased, the pressure inside the can also decreased. When the pressure inside the can decreased, the much higher air pressure outside the can pushed in the sides of the can, causing it to implode.

Figure 16.1

The air molecules in the can hit the inside walls with less energy at low temperature, causing the can to implode as the air temperature decreases.

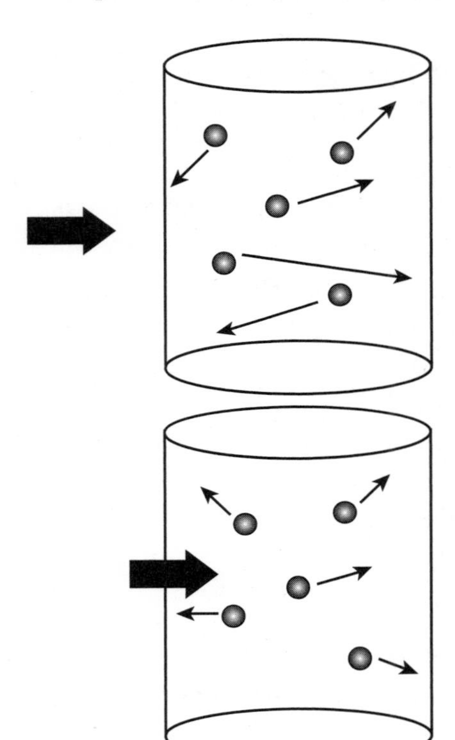

At high temperatures, the force of the gas molecules pushing on the inside of the can is equal to the force of the air pressure pushing on the outside.

At low temperatures, the force of the gas molecules pushing on the inside of the can is less than the force of the air pressure pushing on the outside.

You've Got Problems

Problem 2: I've got a 2.5 L balloon at a temperature of 25° C. If I heat the balloon to a temperature of 50°, what will the new volume of the balloon be?

Gay-Lussac's Law: Spray Paint + Campfire = Bad News

Back in 1802, Joseph Gay-Lussac read a story in the newspaper about a guy who threw a can of spray paint into a campfire. (Editor's note: Only the year in the preceding statement is true.) The unfortunate camper had apparently told his friends that he "wanted to see what would happen" immediately before the accident that claimed his life.

Gay-Lussac knew from the information on the sides of the cans that they shouldn't be stored at high temperatures, but he wanted to know the reason why it was such a bad idea. After a great deal of research, he came up with the following relationship, now known as Gay-Lussac's Law.

$$\frac{P_1}{T_1} = \frac{P_2}{T_2}$$

P_1 is the initial pressure of a gas, T_1 is the initial temperature (in Kelvin), P_2 is the final pressure of a gas, and T_2 is the final temperature in Kelvin. We'll assume that both the volume of the gas and number of moles of gas is constant. If a spray can has an initial pressure of 1.50 atm and an initial temperature of 25° C (298 K), we can compute the internal pressure of the can when the gas inside has reached the temperature of a campfire (500° C, or 773 K).

$$\frac{1.50atm}{298K} = \frac{x}{773K}$$

x = 3.89 atm

From this, Gay-Lussac determined that the increased pressure inside the spray can caused the can to explode in the fire. Because spray cans contain flammable liquids and gases, a huge fireball was created when the can exploded.

The Mole Says

The spray can example in this section assumes that the liquid inside of the can plays an insignificant role in how it behaves while being heated in the campfire. Though the liquid would probably vaporize, we'll assume it doesn't for simplicity's sake.

You've Got Problems

Problem 3: Propane tanks can contain an internal pressure of 35.0 atm before bursting. If the tanks initially hold 20.0 atm of propane at a temperature of 20° C, what is the maximum temperature they can reach before exploding?

The Combined Gas Law

The three gas laws we've mentioned so far in this chapter show that the pressure, volume, and temperature of a gas are all related. Using these equations, we can determine what happens to a gas when we change any of these three variables.

What happens, however, if we want to change two variables at once? Well, we can either use two equations, one after the other, or we can find an equation that includes all three variables. As it turns out, somebody has already done that by formulating the combined gas law. The combined gas law is, straightforwardly enough, a combination of the three laws we just mentioned.

$$\frac{P_1V_1}{T_1} = \frac{P_2V_2}{T_2}$$

To see how this works, let's do a practice problem:

Example: A child at the county fair has let go of her toy balloon. If the initial volume of the balloon was 2.50 L, the atmospheric pressure was 1.00 atm, and the temperature at the fairgrounds was 25° C, what will the volume of the balloon be when it reaches an altitude where the air pressure is 0.250 atm and the temperature is –15° C?

Solution: Simply by plugging these numbers into the combined gas law, we get.

$$\frac{(1.00atm)(2.50L)}{(298K)} = \frac{(0.250atm)(V_2)}{(258K)}$$

$$V_2 = 8.66 \text{ L}$$

You've Got Problems

Problem 4: I've decided to go deep sea exploring in a giant transparent balloon. If I start my voyage in a balloon with a pressure of 1.00 atm, a temperature of 20° C, and a volume of 5.00×10^3 L, what will be the pressure inside the balloon be when I reach a depth where the temperature of the balloon is 2° C and the volume of the balloon is 1,170 L?

Avogadro's Law and the Ideal Gas Law

Let's go back to the Kinetic Molecular Theory for a moment. It states that the molecules in gases are infinitely tiny, and that at any given temperature, all gas molecules have exactly the same amount of kinetic energy. If you'll recall our discussion of rms velocity from Chapter 15, that's why heavy gas molecules travel more slowly than light ones at any temperature.

def•i•ni•tion

The volume of one mole of any gas at standard temperature and pressure is called the **molar volume.**

These properties of a gas lead us to an interesting conclusion. One mole of any gas has exactly the same volume under the same conditions of temperature as one mole of any other gas. The volume of one mole of a gas is called its *molar volume*.

It may not seem immediately obvious why all gases should have the same molar volumes at the same temperatures. Consider this: If the pressure of a gas is equal to the force exerted by gas particles pushing on the sides of whatever container it's stored in, and the volume of a gas depends on its pressure (Boyle's Law), then the molar volumes of every gas are the same. This principle was first understood by Amadeo Avogadro, and is usually referred to as Avogadro's Law.

Since all ideal gases have the same molar volumes, a single equation can be used to express the relationship between the number of moles of a gas present and the volume. This relationship shown below is called the ideal gas law.

$$PV = nRT$$

P denotes pressure (in either atm or kPa), V denotes volume in liters, n is equal to the number of moles of gas, R is the ideal gas constant, and T is the temperature of the gas in Kelvin. There are two possible values for R, 8.314 L kPa/mol K and 0.08206 L atm/mol K. The value used in each problem will depend on the unit of pressure given. For example, if pressure is given in atm, R will be 0.08206 L atm/mol K.

Let's see an example of how this works.

Example: My refrigerator has a volume of 1,100 L. If the temperature inside the refrigerator is 3.0° C and the air pressure is 1.0 atm, how many moles of air are in my refrigerator?

Solution: P = 1.0 atm, V = 1,100 L, R = 0.08206 L atm/mol L (because pressure was given as "atm" in the problem), and T = 276 K. Solve for n using the ideal gas law.

$$(1.0 \text{ atm})(1{,}100 \text{ L}) = n\,(0.08206 \text{ L atm/mol K})(276 \text{ K})$$

$$n = 49 \text{ mol}$$

You've Got Problems

Problem 5: If my oven has a volume of 1,100 L, a temperature of 250° C, and a pressure of 1.0 atm, how many moles of gas does it hold?

Chemistrivia

The ideal gas law explains why hot air balloons work. The number of moles of air inside the balloon will be less than the number of moles of air outside the balloon because the air inside the balloon is warmer than the outside air. Because there are fewer moles of air inside the balloon than outside, the mass of the air in the balloon is also less, causing the balloon to "float" above the surrounding cold air.

Dalton's Law of Partial Pressures

Let's say that, for one reason or another, we're not happy with the regular air we've been breathing our entire lives. Instead of breathing that same old boring air that's floating around outside, we're interested in making "custom air" that fits our youthful, "extreme" personality.

To improve our air, we're going to fill our house with a supercharged air mixture of 40% oxygen by volume, 40% nitrogen by volume, and 20% helium by volume (because high squeaky voices are fun).

As it turns out, John Dalton was also interested in making his own special blend of custom air (Editor's note: This is yet another lie). He reasoned that the total pressure of the custom air in his house would be equal to the sum of the individual pressures of each gas inside the house. His reasoning has been immortalized as Dalton's Law of Partial Pressures, which states:

$$P_{tot} = P_1 + P_2 + P_3 + \dots$$

P_{tot} is the total pressure of all the gases in the mixture, P_1 is the amount of pressure due to gas #1, P_2 is the amount of pressure due to gas #2, and so on. The pressures on the right side of the equation are referred to as *partial pressures* because they represent the pressure that each gas would exert under the same conditions of temperature and volume if the other gases weren't present.

> **def•i•ni•tion**
>
> The **partial pressure** of one gas in a mixture of gases is equal to the amount of pressure that would be exerted by that gas alone if all of the other gases were removed.

As a result, if we decided to pump all of the air out of my house and insert a mixture of air containing 0.300 atm of oxygen, 0.300 atm of nitrogen, and 0.150 atm of helium, the total pressure of the mixture of gases would be:

$$P_{tot} = 0.300 \text{ atm} + 0.300 \text{ atm} + 0.150 \text{ atm}$$

$$P_{tot} = 0.750 \text{ atm.}$$

Because each of the individual gases in a mixture of gases is assumed to be an ideal gas, we can treat each of them independently. As a result, if we only know the number of grams or moles of each gas in the mixture, we can use any of the gas laws discussed earlier in this chapter to find the total pressure of the entire mixture of gases in a problem.

Example: Without doing any prior calculations to see if it's a good idea, I've placed 150 mol O_2, 250 mol N_2, and 75 mol He in my bedroom (which has a volume of 48,000 L and from which I previously removed all the air). If the temperature is 25° C, what is the overall gas pressure inside my bedroom?

Solution: Because each of the gases in this mixture is ideal gas, we can treat each one of them individually using the ideal gas law.

- The partial pressure of O_2:

  $(P_{oxygen})(48{,}000 \text{ L}) = (150 \text{ mol})(0.08206 \text{ L atm/mol K})(298 \text{ K})$

  $P_{oxygen} = 0.076 \text{ atm}$

- The partial pressure of N_2:

  $(P_{nitrogen})(48{,}000 \text{ L}) = (250 \text{ mol})(0.08206 \text{ L atm/mol K})(298 \text{ K})$

  $P_{nitrogen} = 0.13 \text{ atm}$

- The partial pressure of He:

  $(P_{helium})(48{,}000 \text{ L}) = (75 \text{ mol})(0.08206 \text{ L atm/mol K})(298 \text{ K})$

  $P_{helium} = 0.038 \text{ atm}$

Using Dalton's Law, the total pressure of all the gases in this mixture is:

$P_{tot} = P_{oxygen} + P_{nitrogen} + P_{helium}$

$P_{tot} = 0.076 \text{ atm} + 0.13 \text{ atm} + 0.038 \text{ atm}$

$P_{tot} = 0.24 \text{ atm}.$

This is roughly the same air pressure that exists at the top of Mt. Everest, which makes breathing difficult.

You've Got Problems

Problem 6: I've found a pressurized cylinder of nitrogen and oxygen in my basement. The temperature of the cylinder is 15° C and the volume is 55 liters. The partial pressure of oxygen is 5.0 atm and the partial pressure of nitrogen is 8.0 atm. Given this information:

a) How many moles of oxygen and nitrogen are present in the cylinder?

b) What is the total pressure of the gas in the cylinder?

The Least You Need to Know

- Boyle's law, Charles's law, Gay-Lussac's law, and the combined gas law all express the relationships between the volume, temperature, and pressure of a gas.
- The ideal gas law allows us to determine the relationship between pressure, volume, temperature, and number of moles of a gas.
- Dalton's law of partial pressures is good for calculating the pressures within a mixture of gases.

Chapter 17

Phase Diagrams and Changes of State

In This Chapter

- How the vapor pressure of a liquid affects phase changes
- Colligative properties
- What happens when compounds undergo a phase change
- Phase diagrams

In the past five chapters, we've discussed the properties, structures, and behaviors of solids, liquids, and gases. By now you should be a real pro when it comes to the three states of matter.

What we haven't yet mentioned is that we can change materials from one phase to another. The most familiar example is that of water—if you start out with ice, you can heat it until it melts to make liquid water. By additionally heating it to 100° C, we form gaseous water (steam).

In this chapter, we're going to wrap up our explanation of the states of matter by describing how to convert between the phases and how we can manipulate these changes by altering the conditions under which the material exists. By the end of this chapter, you'll not only know about each state of matter, you'll also understand how to control their every action. It's a big responsibility, but I'm sure you're up to it!

Vapor Pressure: Why Phase Changes Occur

When I was a kid, I had a pet fish. I'm sorry to say I didn't do a very good job taking care of it. The fish eventually died because the bowl went dry when I forgot to refill the water. As a budding young scientist, I wondered why the water in the fish tank evaporated. Unfortunately, back at six years old, I wasn't very bright—I concluded that my brother drained the water as a mean trick.

Now that I've been doing chemistry for a while, I realize that the real reason behind the vanishing water was that it evaporated and became a gas.

Why did this happen? Let's consider what happens in a liquid. In any liquid, some molecules have more energy than others. Because some molecules have enough energy to overcome the attractive intermolecular forces between them, they evaporate and enter the gas phase.

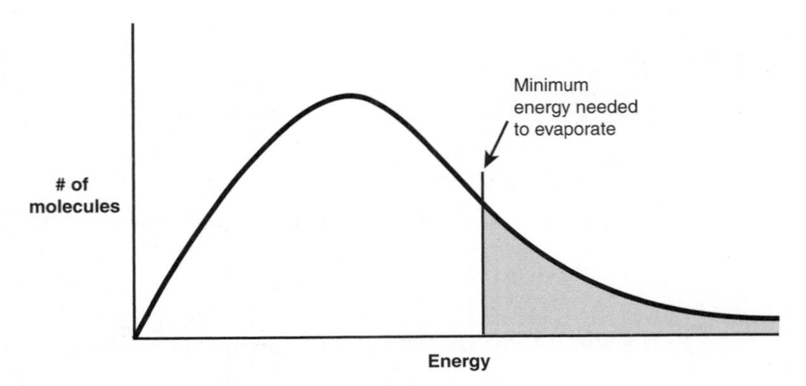

Figure 17.1

Some molecules in a liquid have enough energy to overcome intermolecular forces, so they evaporate.

As the temperature of the liquid increases, the energy of the molecules in the liquid also increases. While some molecules still have more energy than others, more of the molecules have enough energy to evaporate.

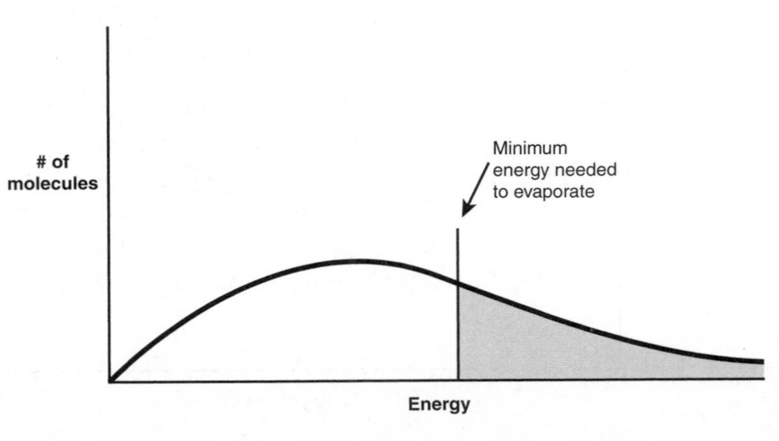

Figure 17.2

At elevated temperatures, larger numbers of molecules have enough energy to evaporate. As a result, evaporation takes place more quickly in warm liquids than cool ones.

Like any other gas, the molecules that have evaporated from the liquid exert a pressure. The pressure exerted by these gas molecules is called the *vapor pressure* of the liquid. The vapor pressure of any pure material or compound, in any state, is dependent only on temperature.

Chemistrivia

You may have already seen how water molecules evaporate more quickly at high temperatures than at low temperatures. When you take a cold shower, your bathroom mirror won't fog because there is very little water vapor in the air. However, if you take a hot shower, more of the water molecules evaporate, causing the cold mirror to fog.

Vapor Pressure and Boiling

As the temperature of a liquid increases, the vapor pressure, due to the evaporation of the molecules, increases. Eventually, when the vapor pressure of the liquid becomes equal to the vapor pressure of the surrounding gas, it begins to boil. The *normal boiling point* of a liquid is defined as the temperature at which its vapor pressure equals one atmosphere.

def•i•ni•tion

The **vapor pressure** of a liquid is the gas pressure in a closed container due to the molecules that have evaporated from the liquid. **Normal boiling point** is the temperature at which a liquid boils at a pressure of 1.00 atm.

From the following figure, we can see the dependence of the vapor pressure of two liquids on temperature.

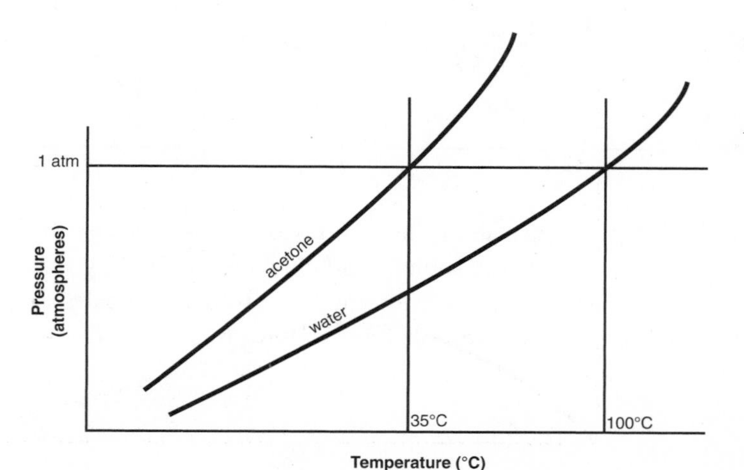

Figure 17.3

The dependence of the vapor pressures of water and acetone on temperature.

> **Chemistrivia**
>
> Liquids with high vapor pressures are said to be "volatile" and those with low vapor pressures are "nonvolatile." Typically, volatile liquids such as gasoline have stronger smells than nonvolatile liquids such as corn syrup because more of the molecules from the volatile liquid are in the gas phase.

As you can see, the vapor pressure of acetone is more than that of water at any given temperature. The reason for this is that the dipole-dipole forces holding acetone molecules together are weaker than the hydrogen bonds holding water molecules together. Consequently, at any given temperature, more acetone molecules than water molecules have enough energy to overcome the intermolecular forces present and enter the vapor phase.

> **You've Got Problems**
>
> Problem 1: Which has a higher vapor pressure, iced tea or rubbing alcohol?

Vapor Pressure and Colligative Properties

The vapor pressure of solutions differs from that of the pure solvents. We can explain this by comparing a pure liquid to a salt water solution.

Figure 17.4

(a) Pure water. (b) Salt water. The larger circles represent the nonvolatile sodium and chloride ions.

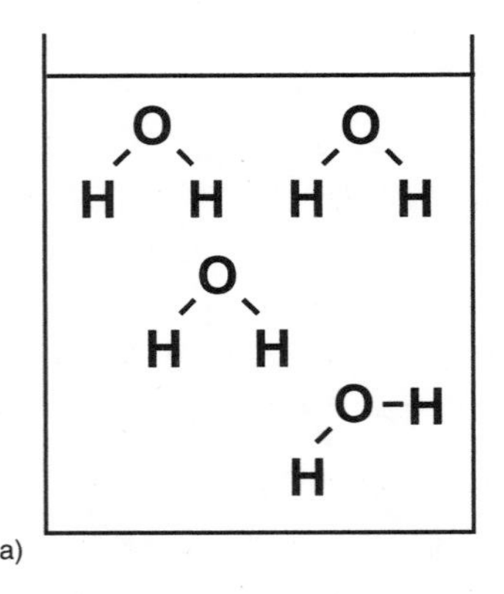

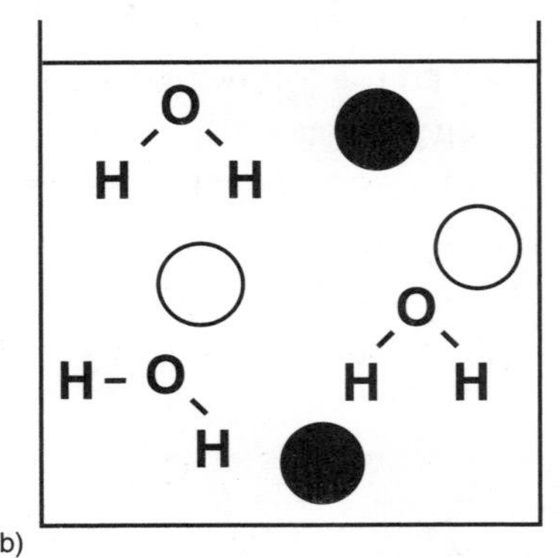

In pure water, any molecule at the surface of the liquid with enough energy to evaporate can do so. However, in salt water, the sodium and chloride ion occupy some of the surface of the liquid, decreasing the area over which the water molecules can evaporate. As a result, the vapor pressure of salt water solutions is less than that of pure water at a given temperature.

Because salt water has a lower vapor pressure than pure water, the boiling point of the water increases. We can see this in the following figure, which shows the dependence of the vapor pressure of both pure and salt water on temperature.

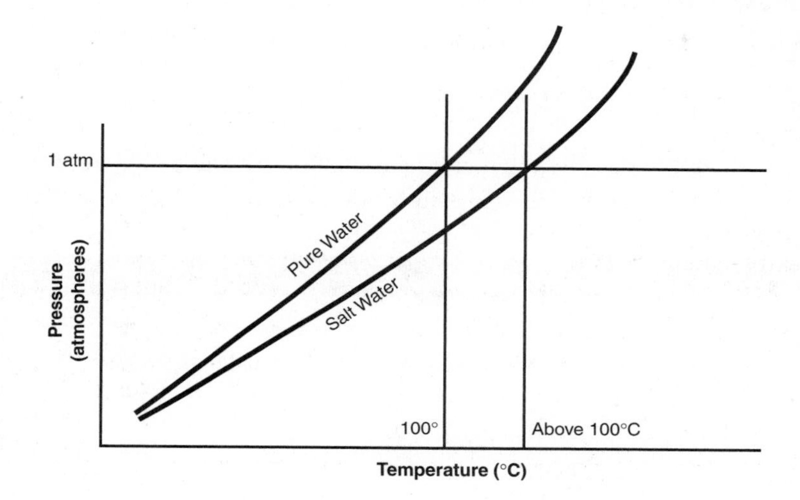

Figure 17.5

The dependence of the vapor pressure of pure and salt water on temperature.

From the diagram, we can see that the temperature required for salt water to boil (which is, again, the temperature at which the vapor pressure is 1 atm) is higher than that required for pure water. This temperature can be calculated from the concentration of the solution using the following equation.

$$\Delta T = K_b m_{solute}$$

ΔT represents the change in boiling point from the pure solvent, K_b is the boiling point elevation constant (which is different for every liquid), and m_{solute} is the molality of the solution (molality, in case you forgot, is moles of solute per kilogram of solvent). The boiling point elevation constant is different for every liquid—for water, it is 0.51 °C/m.

The Mole Says

When finding the molality for the boiling point elevation, we need to find the molality of number of particles in the solution, not just the molality of the compound. For example, if we dissolved 1 mole of NaCl in 1 kg of water, the molality for the purposes of this equation would be 2 m because two ions (the Na^+ and Cl^- ions) are formed when sodium chloride is dissolved.

Example: I like my fruit punch very, very sweet! If I make fruit punch by dissolving 2.0 moles of sucrose in 1,100 grams of water (and adding red food coloring, which we'll ignore for this problem), what will the boiling point of the resulting beverage be?

Solution: The molality of the beverage is found by dividing the moles of solute by the kilograms of solution, or 2.0 mol/1.1 kg = 1.8 m. Plugging this value of molality into the equation with the boiling point elevation constant for water, 0.51° C kg/mol, we get.

You've Got Problems

Problem 2: If I prepare a solution by dissolving 2.5 moles of $ZnCl_2$ in 2.0 kg of water, what will the new boiling point of the solution be?

$\Delta T = (0.051°C/m)(1.8\ m)$

$\Delta T = 0.092°C.$

Because the increase in the boiling point is 0.092°, the overall boiling point of the solution will be 100.092° C.

Chemistrivia

The reason people salt roads in the winter is that the salt forms a concentrated solution with the water from the snow and ice. As salt solutions have lower melting points than pure water, this causes the ice to melt, making it less likely that an unlucky motorist will drive his or her car into a snow bank.

Let's say that we're trying to freeze a salt water solution. Because it's harder for water to solidify when there are sodium and chloride ions present (they get in the way of the attractive forces between them), we need to cool the salt water to a lower temperature before it will begin to freeze. To determine the new melting point of such a solution, we use the following equation.

$$\Delta T = K_f m_{solute}$$

def•i•ni•tion

Colligative properties are any properties of a solution that depend on its concentration.

Where ΔT corresponds to the freezing point depression, K_f is the freezing point depression constant (which is different for each substance), and m_{solute} is the molality of the solute. For water, K_f is 1.86° C/m.

You've Got Problems

Problem 3: What will the concentration of an aqueous NaOH solution have to be to give it a melting point of −1.50° C?

Boiling point elevation and freezing point depression are both called *colligative properties*. Colligative properties are any properties of a solution that depend on the concentration of the solution. Because the melting and boiling point changes both depend on the molality of the solute present, both are colligative properties.

Melting and Freezing

Melting is the process by which a solid becomes a liquid. Freezing is when a liquid is converted to a solid. The temperature at which a material melts is the same at which it freezes because each process is the exact opposite of the other.

What Happens When Something Melts?

You're familiar with what happens when ice melts. As the ice warms up, it starts forming a big puddle until the ice has completely vanished. This is what we can observe with our naked eyes, but what happens at the microscopic level?

The answer is: that depends on the type of material that's melting. As it turns out, ionic and covalent compounds melt in different ways.

For example, let's consider what happens when an ionic compound melts:

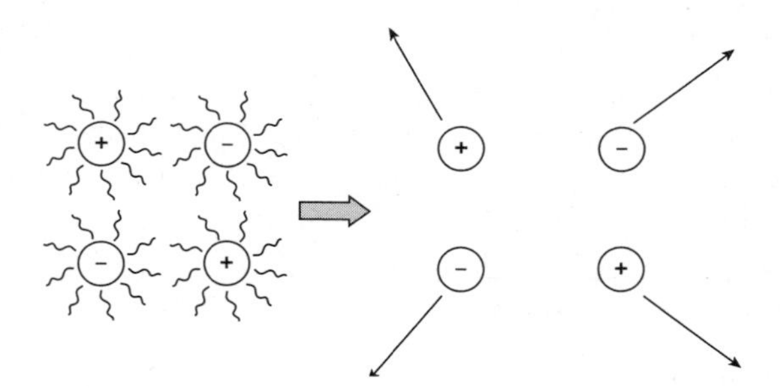

Figure 17.6

Ionic compounds melt when the wiggling of the anions and cations becomes greater than the attraction between them.

In this diagram, we can see what happens when an ionic compound melts. In a crystal, the ions naturally tend to wiggle in place a very small amount. As the temperature of the crystal increases, the ions wiggle more and more, until they reach a point where the ions are wiggling with more energy than the amount of energy holding them together. When this occurs, the ionic compound melts, allowing the ions to move freely throughout the liquid.

In Chapter 8, we learned that the attraction between the cations and anions in an ionic compound is very strong. As a result, ionic compounds generally need to be heated to very high temperatures before they melt.

The Mole Says

About melting in covalent compounds: some students mistakenly believe that covalent bonds are broken when covalent compounds melt. Always remember that it's not the bonds within molecules that are broken; it's the intermolecular forces between molecules that are overcome.

Something similar happens when covalent molecules melt. In a covalent solid, the molecules also vibrate around. When the temperature rises, these molecules gain enough energy that they overcome the strength of the intermolecular forces holding them to each other. This causes the compound to melt and the molecules to move with relative freedom. Because no chemical bonds or strong interactions are broken, covalent compounds require much less energy to melt than ionic compounds.

What Happens When Something Freezes?

When something freezes, the same thing happens as when it melts, except in reverse. As the temperature in a compound decreases, the molecules or ions (depending on the type of compound) have less and less energy. Eventually, the particles have so little energy that the forces between them lock the particles in place, causing the crystal to be reformed.

Boiling and Condensing

Boiling—also called "vaporizing"—is when a liquid is converted to a gas. Condensing—the opposite process—is when a gas is converted to a liquid. Like melting and freezing, these processes are the opposite of one another.

What Happens When a Liquid Boils?

We all know that if you heat a pot of water on the stove, eventually the water turns to steam and vanishes. However, you may not be familiar with why this happens.

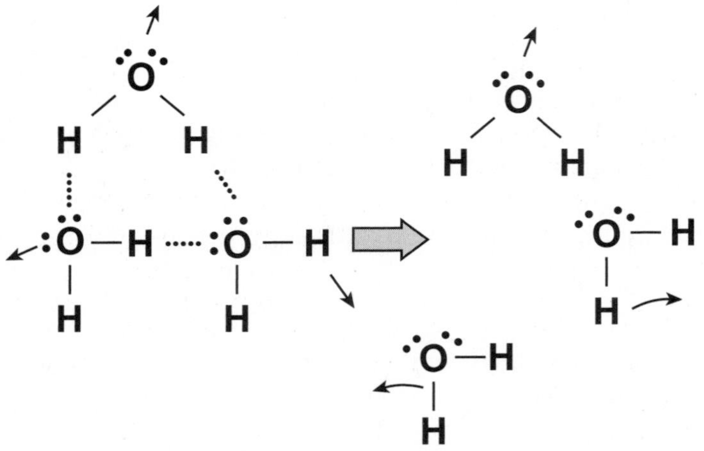

Figure 17.7

Water molecules boil when the amount of energy added to them becomes greater than the hydrogen bonds holding water molecules to each other.

In the preceding figure, we see what happens when water is boiled. Normally, the molecules in liquid water are attracted to one another by hydrogen bonds. When heat is added to water, the molecules move more and more quickly, disrupting the hydrogen bonds that hold them together. Eventually, when enough energy is added, the motion of the water molecules causes them to fly apart and form a gas.

What Happens When a Gas Condenses?

When a gas condenses back into a liquid, the process is exactly the opposite as that of boiling. At some point as a gas is cooled, the energy that the molecules have becomes less than the attractive intermolecular forces between the particles in a gas. When this occurs, the gas condenses back into a liquid.

Sublimation and Deposition

Sublimation is the process by which a solid turns directly into a gas. As you know, in most materials, a solid melts before turning into a gas. However, for some materials, the intermediate liquid phase is bypassed, leading directly to the gas. The reverse process is called deposition, which occurs when a gas is converted directly into a solid.

If you've ever worked with dry ice, you're familiar with the sublimation process. Dry ice is made of solid carbon dioxide, and it goes directly from the solid to the gas phase. As a result, a brick of dry ice seems to vanish over time if placed in a warm room.

Phase Diagrams

So far in this chapter, we've seen how pressure and temperature both affect the state at which matter exists. The big question, then, is how can we express this information in a handy and easy-to-understand diagram?

I'm glad you asked! Some wonderful person has already done this, coming up with something called a "phase diagram." Phase diagrams are neat because they show the phases of a material under all possible conditions of pressure and temperature. To read a phase diagram, find the conditions of pressure and temperature that you're interested in investigating on the chart. The region where this point can be found on the graph will indicate the stable phase of matter for the substance. The phase diagram for water is shown in the following figure.

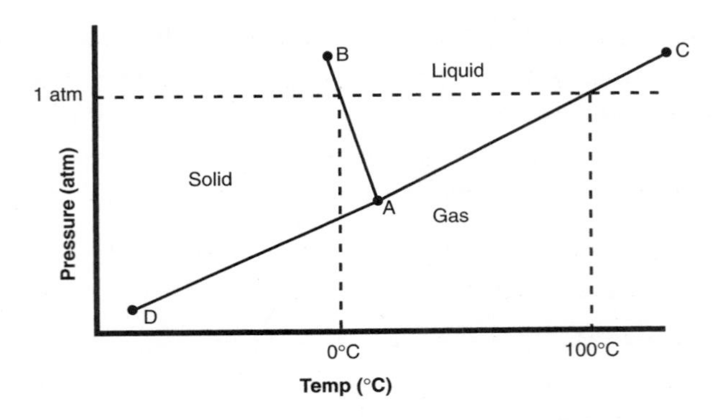

Figure 17.8

The phase diagram of water.

> **Chemistrivia**
>
> If you've got a very cold refrigerator, you can re-create the conditions where solid ice and liquid water are in equilibrium. Make a glass of ice water and place it into the refrigerator before you go to bed. When you wake up in the morning, open the refrigerator—if your refrigerator is exactly at the freezing point, you'll have exactly the same ratio of ice to water in the glass that you started with!

We'll use this chart to learn about the important parts of phase diagrams.

- Point A is called the "triple point." The triple point is the temperature and pressure at which liquid water, water vapor, and solid ice are all stable. As you can see from this diagram, the reason you probably haven't seen water under these conditions is that it corresponds to a very low pressure and temperature.
- The line between point A and point B corresponds to a series of pressure and temperature values where both solid and liquid water can exist in equilibrium at the same time. If you follow the line at a pressure of 1 atm across the table, you can see that the temperature at which you cross this line is 0° C. By heating ice above this temperature, you will cause it to melt; by cooling liquid water below this temperature, you will cause it to boil.
- The line between point A and point C represents the values of pressure and temperature at which the liquid and gas phases are in equilibrium. If you follow the line at 1 atm across the chart, you can see that it tells you that water will boil (go from the liquid to gas phase) at exactly 100° C. Likewise, by cooling steam below a temperature of 100° C, you can cause it to condense.
- The line between point A and point D represents the values of pressure and temperature at which the solid and gas phases are in equilibrium. By changing the conditions of pressure and temperature so water moves directly into the gas

phase without liquefying, you can cause ice to sublime. Likewise, by moving across the line in the other direction, you cause water vapor to be deposited in the solid state.

- Point C is called the critical point. Past the temperature and pressure of the critical point (called the "critical temperature" and "critical pressure"), the liquid and gas phases of water are indistinguishable from each other and exist in an unusual state called a "supercritical fluid." Supercritical fluids can be thought of as gases that have been squished down to the point where the molecules are very, very close to one another so they interact strongly with one another. As a result, supercritical fluids don't behave like normal gases or like liquids, but may have properties of both.

Chemistrivia

Supercritical carbon dioxide is increasingly being used to replace the toxic organic solvents currently used in dry cleaning. By using supercritical CO_2, the clothes get just as clean, but dry cleaners can simply release the gas into the atmosphere when they're finished. Though carbon dioxide is a greenhouse gas, it's much less toxic than dry cleaning solvents!

Phase diagrams are extremely handy because they allow us to figure out what will happen to a state of a material if we change either the pressure or temperature. For example, if you look at the phase diagram for water, you can see that if you have ice at a pressure of exactly 1.00 atm and 0° C, you can cause it to melt by increasing the pressure. You can see this at home by poking a needle firmly into a piece of ice. Because all of the force is concentrated in the very tiny point of the needle, the ice will melt.

You've Got Problems

Problem 4: I have a block of ice at a temperature of −10° C and a pressure of 0.9 atm. Using the phase diagram, determine what I will need to do in order to make this block of ice sublime.

The Least You Need to Know

- Phase changes occur when heat is added or subtracted from a material until the vapor pressure of the initial phase is the same as the vapor pressure of the final phase.
- Solutions of nonvolatile solutes have higher boiling and lower melting points than pure solvents because their vapor pressures are lowered.

- When compounds melt, the molecules have enough energy that they can wiggle away from each other and move freely as a liquid. When compounds boil, the molecules in the liquid have enough energy to completely break free from each other.
- Phase diagrams allow us to keep track of the state of a material under all conditions of temperature and pressure.

Part 4
How Reactions Occur

Chemicals react with each other! I know that probably doesn't come as much of a surprise, but up to this point, we haven't really talked about chemical reactions much.

As it turns out, chemical reactions are a lot like cooking. After all, after mixing a bunch of things together, we end up with something entirely different. We'll start this section by learning about chemical equations and stoichiometry so we have some idea about how we can start cooking in a chemical way.

Once we know how to cook, it's time to learn about how we, as good cooks, can determine how long it will take for our recipes to be complete using the magic of "kinetics." Finally, we'll talk about equilibria—unlike regular cooking, sometimes our finished product turns itself right back into the ingredients!

Chapter 18

Chemical Equations

In This Chapter

- My famous chili recipe
- Balancing chemical equations
- Interpreting the symbols used in chemical equations
- The six types of chemical reaction
- Predicting the products of a chemical reaction

Let's recap everything we've learned about chemistry. We understand how to use the scientific method to solve problems. We understand the bonding and properties of ionic and covalent compounds. We've become experts where the periodic table is concerned. We can predict how solids, liquids, and gases will behave when we experiment with them.

Though we know a lot of chemistry, we haven't talked about chemical reactions! Something tells me that you'd like to learn about chemical reactions, so I suppose it's only right that we talk about them. However, before we can specifically talk about various types of chemical reactions, we must first learn the notation and terminology that chemists use to discuss them. I'm speaking, of course, about chemical equations.

Many students think that chemical equations are hard to understand and write because they contain a lot of strange and unfamiliar symbols. If you're one of these people, don't worry about it—as you'll soon see, chemical equations are a piece of cake!

Let's Make Chili!

My family has a long-standing tradition that we follow every New Year's Day. While most people are watching the Rose Bowl parade on television or nursing terrible hangovers, my family is firing up the stove, dicing onions, and yelling at the cat to get off the kitchen counter. You see, we take our New Year's Day chili very seriously.

Every member of my family makes chili differently. Some favor a heavy, stew-like chili, while others prefer a hot, less textured chili. However, in the entire history of my family, nobody has *ever* used a recipe, and nobody has ever told anyone else how their chili is made.

A few years back, I decided to write down my chili recipe so I could let other people enjoy the result of my considerable culinary expertise. Though it may cost me the love and respect of my family, in this chapter you're going to learn how to make chili the Guch way. Here it is!

Ian Guch's New Year's Day Chili With Beans

1 lb. browned hamburger

16 oz. can tomato sauce

2–16 oz. cans of red kidney beans

1 lb. chopped onion

3 chopped green peppers

3 cloves fresh garlic

½ cup chili powder (add more if you like it hot)

2 tbsp. olive oil

Combine the ingredients in a large pot and simmer over low/medium heat for at least two hours (the longer the better). If it seems too watery, uncover to evaporate water until it reaches the desired thickness. If it seems too thick, add an additional 8 oz. tomato sauce.

You may be wondering why I decided to tell you about chili in a chemistry book. It's simple—cooking and chemistry are the same thing, except that in chemistry you usually shouldn't eat whatever it is you make. In this chapter, I'll explain to you how you already understand chemical reactions, though in food form.

Getting Our Ingredients

As in cooking, when we perform chemical changes, a list of ingredients is required. However, the form used to express these ingredients is a bit different. Let's take a look at the list of ingredients needed to make water.

$$H_2 + O_2 \rightleftharpoons H_2O$$

This statement tells us a great deal of information. Everything written before the arrow is required to make the chemical reaction take place and is called a *reactant*. The formulas written after the arrow are the chemicals that are formed and are called the "products" of the reaction. Together, the whole statement is called a chemical equation.

All chemical changes are expressed using equations in this general form. The numbers of reactants and products may change from one equation to another, but the general format is always the same.

def•i•ni•tion

The starting materials for a chemical reaction are referred to as **reactants** or **reagents.** The chemicals that are made in the reaction are referred to as **products.** Together, the entire statement that includes both the products and reactants is called a chemical equation.

Chemistrivia

Most chemistry textbooks write the arrows for chemical reactions like this: $\rightarrow$. However, you'll notice that in this chapter and beyond, I'll write the arrows like this: $\rightleftharpoons$. The reason I do such an unusual thing is that all chemical reactions are reversible, meaning that in addition to reactants forming products, the products can also react in reverse to form reactants. Though in many cases the backwards reaction doesn't occur at a significant rate, it does occur, making it important to write the $\rightleftharpoons$ arrow instead of the $\rightarrow$ arrow. We'll discuss this in greater detail in Chapter 22.

How Much of Each Ingredient Do We Need?

All good recipes tell the chef how much of each ingredient is required. The process by which we figure out how much of each reactant is required for a chemical process is called balancing the equation.

Equation Balancing Made Easy

Let's learn equation balancing by studying the equation for making water from hydrogen and oxygen: $H_2 + O_2 \rightleftharpoons H_2O$. This equation, as well as any others you may come across, can be balanced by following these steps.

Step 1

Before balancing equations, tell yourself never to change any of the chemical formulas of either the reactants or products. If you change the formulas by adding subscripts or altering them in any way, your equation is guaranteed to be wrong!

Step 2

Draw a table that shows the number of atoms of each element both before and after the arrow. For the reaction of hydrogen and oxygen to make water, this table would look like this.

Element	Before the Arrow	After the Arrow
H	2	2
O	2	1

Step 3

Change the equation in a way that will make the number of atoms of each element in the "before" and "after" columns match one another. To do this, the formulas of the chemicals involved can't be changed. Instead, we add numbers called "coefficients" in front of the formulas in the equation. These coefficients allow us to multiply the number of atoms of each element in the compound so the columns will match up.

The Mole Says

The law of conservation of mass makes it important that the number of atoms of each element doesn't change during the equation. If there were more atoms of an element on one side of the equation than the other, this would imply either the formation or destruction of matter, which the law of conservation of mass forbids.

Examining our table, we see that the hydrogen columns already match, so we'll change the equation to make the oxygen columns match. There are two oxygen atoms before the arrow and only one after the arrow, so we'll put a "2" in front of the molecule containing oxygen after the arrow to give us the following equation.

$$H_2 + O_2 \rightleftharpoons 2\ H_2O$$

Step 4

Redo the table to reflect the change in the equation. If the columns match, the equation is balanced and we're done! If the columns don't match, we need to go back to step 3 and change another number.

In our example, we see that the table has changed in the following way.

Element	Before the Arrow	After the Arrow
H	2	4
O	2	2

As you can see, the oxygen columns now match perfectly, which is exactly what we were trying to do by adding the "2" in front of H_2O. Unfortunately, the hydrogen column, which previously had matched, is now unbalanced. Though this may make it seem as if we've made a mistake, this is a common event when balancing equations.

After an examination of the revised table, it becomes clear that we should add a "2" in front of H_2 before the arrow so we'll have four hydrogen atoms before the arrow. Our revised equation now looks like this:

$$2\ H_2 + O_2 \rightleftharpoons 2\ H_2O$$

Because we've changed another number, we can redo our table to find that the numbers of hydrogen and oxygen atoms are the same on both sides of the equation.

Element	Before the Arrow	After the Arrow
H	4	4
O	2	2

Equation Balancing Cautions, Tips, and Suggestions

As with recipes, some equations are more difficult than others. Though these may initially be frustrating, below are some tips you can try when the going gets tough:

- If you've been working on an equation for a few minutes and aren't getting any closer to solving it, start over again from scratch. Though this doesn't guarantee success, it will allow you to get a fresh perspective on what you're doing.
- If you still can't solve the problem, start over and put a "2" in front of the most complicated-looking compound in the equation. If you still can't solve the problem, start the equation over by writing a "3" in front of this molecule. Eventually, you'll be able to solve the problem.

The Mole Says

The best way to learn how to solve equations is to practice doing them over and over. The tricks in this section help me to solve equations; with practice, you'll discover tricks of your own!

- If you can reduce all of the coefficients in an equation by a lowest common denominator, do it! An example of what I mean is shown here.

  $4\ H_2 + 2\ O_2 \rightleftharpoons 4\ H_2O$

If you made the table showing the number of atoms in this equation, it would work out just fine. However, it's more proper to write this in a reduced form by dividing all of the subscripts by two to yield the equation that we solved earlier.

You've Got Problems

Problem 1: Balance the following equations:

a) $CaCl_2 + AgNO_3 \rightleftharpoons AgCl + Ca(NO_3)_2$

b) $(NH_4)_2CO_3 + FeBr_3 \rightleftharpoons Fe_2(CO_3)_3 + NH_4Br$

c) $P_4 + O_2 \rightleftharpoons P_2O_5$

Writing Complete Equations

As with recipes, all chemical equations include practical information for making our product. In the chili recipe, it mentioned that we should "simmer for at least two hours." In chemical equations, we use somewhat different terms.

The following are symbols that are commonly found in chemical equations to indicate the states of the products and reactants, as well as what reaction conditions are required.

Symbols of State

To indicate the states of the products and reactants in a reaction, we write the following symbols as subscripts after each chemical in the equation.

Symbol	What It Means	Example
(s)	The chemical is a solid	$Fe_{(s)}$
(l)	The chemical is a liquid	$H_2O_{(l)}$
(g)	The chemical is a gas	$N_{2(g)}$
(aq)	The chemical is dissolved in water	$AgNO_{3(aq)}$

Sometimes it's easy to tell what symbols of state should be used, and sometimes it's not. For example, water is frequently in a liquid form. However, if we do a chemical reaction in which a large amount of heat is required, it may be a gas (steam).

Reaction Conditions

Frequently, symbols are written around the arrow in a chemical reaction to indicate to the reader what procedures need to be followed to make a chemical reaction occur. Here are some of the most common symbols.

Symbol	What It Means
Δ	Add energy/heat to the reactants.
100° C	Heat the reactants to the specified temperature.
2 atm	The reactants should be combined at the specified pressure.
reflux	Continuously heat the reactants to boiling, and recondense the vapor.
chemical formula	The specified chemical is needed for the reaction to proceed, or is the solvent.
3 hrs	The reaction should proceed for the specified quantity of time.

Arrows

Sometimes you will observe arrows written immediately after the chemical formula of the products in a chemical reaction. An arrow pointing up (as in $CO_2\uparrow$) indicates that the product will form a gas that will bubble out of a solution. An arrow pointing down (as in $PbI_2\downarrow$) indicates that the product will spontaneously precipitate from the solution. ("Precipitate," in this sense, means the formation of a solid from the combination of two aqueous solutions.)

The symbols I've indicated in the previous table are by no means the only ones you'll find in chemical equations; depending on the type of reaction you're performing, there may be others. However, these are the most commonly seen in an introductory chemistry course.

As an example, let's add the appropriate symbols into the equation for the formation of water from hydrogen and oxygen. The reaction proceeds as follows: When energy is added to a mixture of hydrogen and oxygen gases, steam is formed. Using the symbols we discussed, the complete equation for this reaction is the following.

$$2\ H_{2(g)} + O_{2(g)} \overset{\Delta}{\rightleftharpoons} 2\ H_2O_{(g)}$$

You've Got Problems

Problem 2: Write complete chemical equations for the following reactions:

(a) When dissolved lead(II) nitrate is added to an aqueous solution of potassium iodide, lead(II) iodide precipitates from the solution, and dissolved potassium nitrate is formed.

(b) When iron powder is heated in the presence of oxygen gas, iron(III) oxide powder is formed.

Adding Variety to Our Menu

To be an excellent and accomplished cook, knowing how to prepare many different dishes using many different methods is essential. For example, you can't be a good cook if you only know how to work a fryer. To be a good cook, you must also know how to broil, sauté, bake, braise, poach, etc.

Many textbooks and chemistry teachers describe reactions as being one of six or so different types. Though I would argue that chemical reactions are all basically the same (they all make products from reactants), it's usually not taught that way, so I'll very briefly describe the six types of reaction so you'll know what other people are talking about. However, I hope that in your own mind you realize that these types of reaction are arbitrarily assigned and don't really mean much.

- **Combustion reaction.** Combustion reactions occur when organic molecules combine with oxygen to form carbon dioxide, water vapor, and a large quantity of heat. A simple example of such a reaction is the combustion of methane, a constituent of natural gas.

  $$CH_{4(g)} + 2\ O_{2(g)} \overset{\Delta}{\rightleftharpoons} CO_{2(g)} + 2\ H_2O_{(g)}$$

- **Synthesis reaction.** Synthesis reactions are when small molecules combine to form larger ones. A commercially important example of a synthesis reaction occurs during the Haber process, which results in the formation of ammonia from nitrogen and hydrogen.

  $$N_{2(g)} + 3\ H_{2(g)} \overset{\Delta}{\rightleftharpoons} 2\ NH_{3(g)}$$

- **Decomposition reaction.** Decomposition reactions are the opposite of synthesis reactions, and take place when large molecules break apart to form smaller molecules. An example of a decomposition reaction is when carbon dioxide bubbles are formed by the decomposition of carbonic acid in a bottle of soda.

  $$H_2CO_{3(aq)} \rightleftharpoons H_2O_{(l)} + CO_{2(g)}\uparrow$$

- **Single displacement reactions (also called single replacement reactions).** Single displacement reactions occur when a pure element switches places with one of the elements in a chemical compound. An example of this type of reaction occurs when zinc reacts with acetic acid to form hydrogen gas and zinc acetate.

  $Zn_{(s)} + H_2CO_{3(l)} \rightleftharpoons H_{2(g)}\uparrow + ZnCO_{3(aq)}$

- **Double displacement reaction.** These reactions occur when the cations of two ionic compounds switch places. A double displacement reaction takes place when dissolved magnesium sulfate is added to sodium hydroxide.

  $MgSO_{4(aq)} + 2\ NaOH_{(aq)} \rightleftharpoons Mg(OH)_{2(s)}\downarrow + Na_2SO_{4(aq)}$

- **Acid-base reactions.** Acid-base reactions take place when an OH^- and H^+ ion combine to form water. An acid-base reaction takes place when household ammonium hydroxide is added to household vinegar, which contains acetic acid.

  $NH_4OH_{(aq)} + HC_2H_3O_{2(aq)} \rightleftharpoons NH_4C_2H_3O_{2(aq)} + H_2O_{(l)}$

In case all of this isn't entirely clear, there's an easy way to figure out what kind of reaction is taking place. To do this, ask yourself the following questions below in order, from first to last. When you can answer "yes" to any of these questions, you've found the type of reaction taking place! Important: don't go down this list out of order—even though it's very possible that you can answer "yes" to more than one of the questions below, only the first one you can answer "yes" to is correct!

1. Does the equation contain CO_2, H_2O, *and* O_2? If yes, it's a combustion reaction.
2. Does the equation involve less complicated molecules combining to form much larger ones? If yes, it's a synthesis reaction.
3. Does the equation involve complicated molecules breaking down to form less complicated molecules? If yes, it's a decomposition reaction.
4. Are there any elements in the equation that are present by themselves (i.e., not bonded to other elements)? If yes, it's a single displacement reaction.
5. Is water a product of this reaction? If yes, it's an acid-base reaction. If no, it's a double displacement reaction.

Let's see how this works using the example of the reaction of potassium hydroxide with calcium chloride: $2\ KOH + CaCl_2 \rightleftharpoons 2\ KCl + Ca(OH)_2$.

Going down the list, we can see from the first question that the equation contains neither water, carbon dioxide, or oxygen, so it's not a combustion reaction. The compounds on either side of the arrow seem to have about the same complexity, so it's neither a synthesis nor a decomposition reaction. There are no elements that haven't bonded to other elements, so it's not a single displacement reaction. Getting to the final question, we see that water is not a product. As a result, this is a double displacement reaction!

You've Got Problems

Problem 3: Identify the type of reaction taking place in each of the equations below:

a) $AgNO_{3(aq)} + HCl_{(aq)} \rightleftharpoons AgCl_{(s)}\downarrow + HNO_{3(aq)}$

b) $Cu_{(s)} + AgNO_{3(aq)} \rightleftharpoons CuNO_{3(aq)} + Ag_{(s)}$

c) $Pb(OH)_{2(s)} + H_2SO_{4(aq)} \rightleftharpoons PbSO_{4(s)} + 2\ H_2O_{(l)}$

d) $C_2H_{4(g)} + 3\ O_{2(g)} \rightleftharpoons 2\ CO_{2(g)} + 2\ H_2O_{(g)}$

Predicting Reaction Products

When cooking, it's frequently handy to predict what will happen when we mix a bunch of ingredients together. For example, if we're interested in making a delicious new salad dressing, we would have a very small chance of making anything edible if we had no way of knowing which ingredients would have the greatest chance of succeeding.

Likewise, it's often necessary for chemists to predict the chemical reactions that will take place when two chemicals are combined. For example, if we're adding a chemical to a tank of toxic waste to stabilize it, we'd be very unhappy if we failed to predict an explosive reaction.

An easy way to predict what reaction will take place when two chemicals are mixed is to identify the type of reaction that's likely to occur when the chemicals are combined. Of course, we mentioned before that these types of reaction are arbitrary, but they do sometimes have a useful purpose.

The Mole Says

The tips in this section, while helping you to figure out what reaction might occur, aren't infallible in correctly predicting the reaction that will take place. However, if you're not sure what will happen, these tips will be useful in suggesting some possibilities.

Here are some tips you may find handy in helping to predict the type of reaction that will occur if you know only the reactants. Keep in mind that not all combinations of chemicals will result in a chemical reaction—these tips are handy only for helping to predict what would happen should they happen to react.

- If two ionic compounds are combined, it's usually safe to predict that a double displacement reaction will occur.

- If the chemicals mixed are oxygen and something containing carbon, it's usually a combustion reaction.
- If we start with only one reactant, the reaction taking place is probably a de-composition reaction. To predict the products of such a reaction, see what happens if the chemical breaks into smaller, familiar products such as water, carbon dioxide, or any of the gaseous elements.
- When pure elements are combined, synthesis reactions are the frequent result.

Bad Reactions

There are two common mistakes when predicting the products of a chemical reaction. The first is predicting the formation of a theoretically impossible product such as $NaCO_3$ or Ag_4Cl. The second is failing to balance the equation once the products have been accurately predicted.

- If a pure element combines with an ionic compound, a single displacement reaction may take place.
- If a compound containing the hydroxide ion is involved, check the other compound to see if it contains hydrogen. If it does, it may be an acid-base reaction.

You've Got Problems

Problem 4: Write balanced chemical equations for the reactions that might occur when the following reactants are combined:

(a) $NaOH + H_2SO_4 \rightleftharpoons$?

(b) $NH_3 + I_2 \rightleftharpoons$?

(c) $C_3H_8O + O_2 \rightleftharpoons$?

The Least You Need to Know

- Chemical equations are like recipes because both tell you how to produce something desirable from basic ingredients.
- It's important to balance equations in order to follow the law of conservation of mass.
- The states of the reactants and products of a chemical reaction, as well as the actions we need to perform to make a reaction take place, can be expressed as symbols added to the chemical equation.

- Many people claim that there are six types of chemical reaction, each of which has its own properties. In reality, chemical reactions aren't as easy to pin down.
- It's usually possible to predict the products of a chemical reaction by determining the type of chemical reaction that's likely to occur when the reactants are combined.

Chapter 19

Stoichiometry

In This Chapter

- Simple stoichiometry calculations
- Limiting reactant problems
- Gas stoichiometry

As you may have gathered from the previous chapter, I like to cook. My chili recipe is perfect for serving a group of 10 (or a group of 4, with lots and lots of leftovers). However, what would happen if instead of one pound of hamburger, I had only half a pound? What would become of the recipe then?

You're probably thinking to yourself, "What a bonehead! Just scale back the recipe so that the quantities of the other ingredients are also halved!" If that statement makes sense, you already understand everything that's contained in this chapter. Unfortunately, chemists like to use big fancy words when describing chemical reactions, so you'll still have to read the rest of the chapter. Throughout the course of this chapter, remember that you already understand this material—it's just a matter of translating what you already know into chemical terms.

Stoichiometry: Fun to Say, Fun to Do!

Before I write another paragraph, let's all pronounce the word "stoichiometry" together. Ready, set, "stoy-key-ah-meh-tree." Say it again! Now, say it five times as quickly as you can. I told you it was fun to pronounce!

Now for the hard question: What does it mean? *Stoichiometry* is simply a way of relating the masses or volumes of products and reactants in a chemical reaction to each other. Put in a simpler way, it's how we can figure out how much of each ingredient will be needed to make a desired quantity of a final product.

> **def•i•ni•tion**
>
> **Stoichiometry** is a way of relating the masses or volumes of the reactants and products of a chemical reaction to each other.

To illustrate what I mean, let's use another recipe that my wife is fond of (she's not as good a cook as I am, so the recipe is much simpler).

Mrs. Guch's Old-Fashioned Ice Water Recipe

1 glass of water

4 ice cubes

Place ice cubes into water. Makes one glass of ice water.

I told you she wasn't much of a cook. In any case, to prove that you already know stoichiometry, I want you to calculate how many glasses of ice water can be made if I have 5 glasses of water and an *excess quantity* of ice.

> **def•i•ni•tion**
>
> The term **excess quantity** is common in stoichiometric calculations. This term means that we have a larger than needed amount of the "excess" reactant and a smaller quantity of the other reactant (called the "limiting reactant," but more on that later). As a result, the amount of product that will be formed depends on the other reactant, not the "excess" one.

Okay, time's up. If you determined that we could make five glasses of ice water with the specified ingredients, you're already a stoichiometry genius! If you couldn't, then you should go get yourself five glasses of water and a huge sack of ice and perform this experiment to prove to yourself that five glasses is the correct answer.

Simple Stoichiometry Calculations

Now that you've done your first stoichiometric calculation using the ice water, it's time to move on to the sorts of questions that chemistry teachers like to ask. Here's one now:

Example: Using the equation $2\ H_{2(g)} + O_{2(g)} \rightleftharpoons 2\ H_2O_{(g)}$, determine how many moles of water can be formed if six moles of oxygen are made to react with an excess of hydrogen.

Yikes! That doesn't look much like the ice water example, does it? Before we even attempt to solve this problem, let me give you a handy diagram that will solve all of your stoichiometric needs.

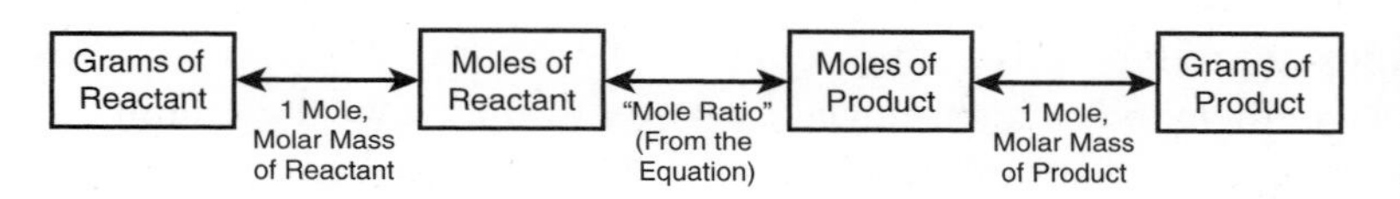

Figure 19.1

This handy chart will help you with all mass-to-mass or mole-to-mole stoichiometry problems, like the one in our example.

To use the diagram, follow these steps:

1. Find the box that corresponds to the information that you were given in the problem. In our example, we were given "6 moles of oxygen," so the "Moles of Reactant" box is where we'll start. (If we were given "6 grams of oxygen" instead, we'd start at the box labeled "Moles of Reactant.")
2. Find the box that corresponds to the value that you're ultimately trying to find. Because our problem wants us to determine "how many moles of water can be formed," we want to end up at the box that corresponds to "Moles of Product."
3. Write down the number and unit that you were given in the problem on your paper.

   6 mol O_2

4. Write a multiplication sign after the number and unit you were given, followed by a straight, horizontal line.

   6 mol O_2 × ________________

5. Below the line, write the same unit that you were given. Include any chemical formulas.

   $$6 \text{ mol } O_2 \times \frac{\qquad\qquad}{\text{mol } O_2}$$

The Mole Says

Stoichiometric calculations are done in the same manner as the unit conversions in Chapter 2 or the mole calculations in Chapter 11. If you have problems with stoichiometry, it might not be a bad idea to go back and do some unit conversion problems as a warm-up.

6. The diagram I gave you is a map that tells you how to get to where you're going. From the box that corresponds to the information you were given in the problem, move one step toward your destination. Write the unit from this destination box above the line. In our example, since the next box says "Moles of Product," we'll write "mol H_2O" above the line.

$$6 \text{ mol } O_2 \times \frac{\text{mol } H_2O}{\text{mol } O_2}$$

7. From the line in between the boxes in the diagram, get the conversion factors for this calculation, and put them in front of the appropriate unit. In this example, our conversion factor is something called the "mole ratio," which consists of the ratio of moles of product to the ratio of moles of reactant. In our mole ratio step, we use the coefficients in the balanced equation to find what these numbers are. In our case, the equation for this reaction has a "2" in front of water and a "1" in front of oxygen, so use these numbers as your mole ratio.

$$6 \cancel{\text{mol } O_2} \times \frac{2 \text{ mol } H_2O}{1 \cancel{\text{mol } O_2}}$$

8. Return to step 4 of this example, and continue the process until you have reached the destination box. In our case, the second box is the same as the destination box, so we don't need to continue. However, if we were trying to find "grams of water," we'd need to add another step to our conversion.

9. Once the calculation has been completely set up, solve the resulting equation to get the final answer. The unit in our answer will be the only one that doesn't cancel out—in this case, "mol H_2O."

$$6 \cancel{\text{mol } O_2} \times \frac{2 \text{ mol } H_2O}{1 \cancel{\text{mol } O_2}} = 12 \text{ mol } H_2O$$

The Mole Says

Though we did our stoichiometry diagram to find the number of moles of product that can be found from a given quantity of reactant, we can just as easily use it to determine the number of moles of reactant that will be required to produce a given quantity of product. In such an example, we'd simply go from right to left in the diagram, instead of left to right. The math is done in *exactly* the same way!

You've Got Problems

Problem 1: Using the following equation, determine how many moles of calcium hydroxide can be made from 125 grams of calcium chloride:

$CaCl_{2(aq)} + 2\ NaOH_{(aq)} \rightleftharpoons Ca(OH)_{2(s)} + 2\ NaCl_{(aq)}$

Limiting Reactant Problems

Now that you're a pro at simple stoichiometry problems, let's try a more complex one.

Using the recipe for ice water (1 glass of water + 4 ice cubes = 1 glass of ice water), determine how much ice water we can make if we have 10 glasses of water and 20 ice cubes.

Hopefully, you didn't have too much trouble figuring out that we can make only five glasses of ice water. Let's go through this calculation carefully to see what we did (it'll be clear why we need to do this in a second).

- Using our recipe, we can make 10 glasses of ice water with 10 glasses of water.
- With the same recipe, we can make 5 glasses of ice water with 20 cubes of ice.
- Because we run out of ice before we run out of water, we can only make five glasses of ice water. There will be five glasses of warm water left over.

def•i•ni•tion

The **limiting reactant** in a stoichiometry problem is the one that runs out first, which limits the amount of product that can be formed. The other reactant is called the **excess reactant.**

In our example, we would say that ice is the *limiting reactant*. The ice is said to be "limiting" because it is the ingredient we would run out of first, which puts a limit on how much ice water we can make. The water is called the *excess reactant* because we had more of it than was needed.

Limiting Reactants in Chemistry

We can use this method in stoichiometry calculations. Again, if we're given a problem where we know the quantities of both reactants, all we need to do is figure out how much product will be formed from each. The smaller of these quantities will be the amount we can actually form. The reactant that resulted in the smallest amount of product is the limiting reactant. Let's see an example.

Example: Using the equation 2 $H_{2(g)}$ + $O_{2(g)}$ $\leftrightarrow$ 2 $H_2O_{(g)}$, determine how many moles of water can be formed if I start with 1.75 moles of oxygen and 2.75 moles of hydrogen.

Solution: Do two stoichiometry calculations of the same sort we learned earlier. The first stoichiometry calculation will be performed using "1.75 mol O_2" as our starting point, and the second will be performed using "2.75 mol H_2" as our starting point.

$$1.75\ \cancel{\text{mol O}_2} \times \frac{2\ \text{mol H}_2\text{O}}{1\ \cancel{\text{mol O}_2}} = 3.50\ \text{mol H}_2\text{O}$$

$$2.75\ \cancel{\text{mol H}_2} \times \frac{2\ \text{mol H}_2\text{O}}{2\ \cancel{\text{mol H}_2}} = 2.75\ \text{mol H}_2\text{O}$$

Because "2.75 mol O_2" is the smaller of these two answers, it is the amount of water that we can actually make. The limiting reactant is hydrogen because it is the reactant that limits the amount of water that can be formed since there is less of it than oxygen.

You've Got Problems

Problem 2: Using the following equation, determine how much lead iodide can be formed from 115 grams of lead nitrate and 265 grams of potassium iodide:

$Pb(NO_3)_{2(aq)}$ + 2 $KI_{(aq)}$ $\rightleftharpoons$ $PbI_{2(s)}$ + 2 $KNO_{3(aq)}$

How Much Excess Reactant Is Left Over?

Once we've determined how much of each product can be formed, it's sometimes handy to figure out how much of the excess reactant is left over. This task can be accomplished by using the following formula.

Figure 19.2

Formula for finding out how much excess reactant is left over.

$$\text{Amount of Excess Reactant Left Over} = \text{Original Quantity of Excess Reactant} - \left[\text{Original Quantity of Excess Reactant}\left(\frac{\text{Amount of product predicted by the limiting reactant}}{\text{Amount of product predicted by the excess reactant}}\right)\right]$$

In our limiting reactant example for the formation of water, we found that we can form 2.75 moles of water by combining part of 1.75 moles of oxygen with 2.75 moles of hydrogen. Because hydrogen was the limiting reactant, let's see how much oxygen was left over.

O_2 = 1.75 mol O_2 – (1.75 mol O_2)(2.75 mol/3.50 mol) remaining

= 0.375 mol O_2 remaining

You've Got Problems

Problem 3: Using your results from problem #2 in this chapter, determine the amount of excess reactant left over from the reaction.

Gas Stoichiometry

Though we've limited our discussion of stoichiometry to grams and moles, we can also do stoichiometric calculations for gases using volume. However, in order to do this, we need to modify our diagram slightly.

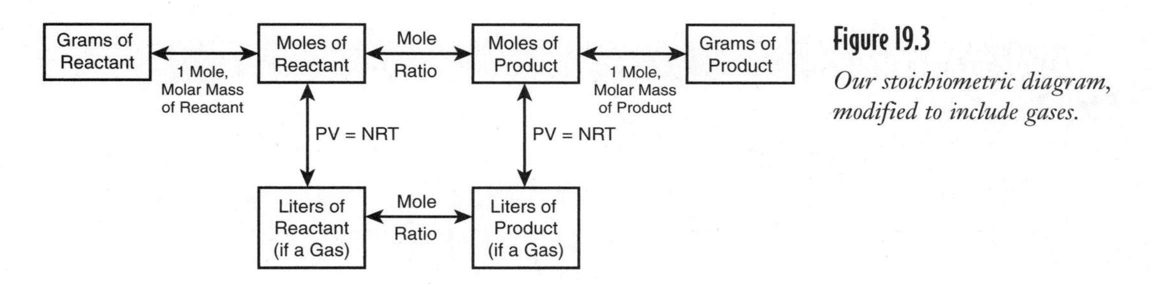

Figure 19.3

Our stoichiometric diagram, modified to include gases.

In order to use this diagram, we need to be able to convert from liters of a gas to moles. Fortunately, we learned how to do this in Chapter 16 with the ideal gas law. If you've forgotten how to use the ideal gas law, it might be a good idea to brush up on it before continuing with this section!

Aside from this change, stoichiometric calculations for gases are done in exactly the same way. Start at the box that includes the information you've been given, and move through the diagram, box by box, until you arrive at your desired destination. Let's do an example:

Example: For the reaction 2 $H_{2(g)}$ + $O_{2(g)}$ $\rightleftharpoons$ 2 $H_2O_{(g)}$, determine how many liters of hydrogen gas will be required to produce 175 grams of water vapor (steam). Assume that you have an excess of oxygen gas, a partial pressure of hydrogen of 1.00 atm, and a temperature of 20° C.

Solution: This problem is solved in exactly the same way as the other stoichiometry problems in this chapter. In order, we'll need to convert the number of grams of steam to moles of water, then moles of water to moles of hydrogen, and finally the moles of hydrogen to liters of hydrogen.

Step 1

Convert grams of steam to moles of water:

$$175 \not{g\,H_2O} \times \frac{1 \text{ mol } H_2O}{18.0 \not{g\,H_2O}}$$

Step 2

Convert moles of water to moles of hydrogen:

$$175 \not{g\,H_2O} \times \frac{1 \not{mol\,H_2O}}{18.0 \not{g\,H_2O}} \times \frac{2 \text{ mol } H_2}{2 \not{mol\,H_2O}} = 9.72 \text{ mol } H_2$$

Step 3

Convert moles of hydrogen to liters of hydrogen using the ideal gas law, PV=nRT. In this example, P = 1.00 atm, V is unknown, n = 9.72 mol, R = 0.08206 L atm/mol K, and T = 293 K (remember always to convert degrees Celsius to Kelvin when doing gas law problems).

$$(1.00atm)(V) = (9.72mol)(0.08206\frac{Latm}{molK})(293K)$$

V = 234 L

The Least You Need to Know

- Stoichiometry is the mathematical method by which we relate the quantities of products and reactants of a chemical reaction to one another.
- The reactant that runs out first in a chemical process is called the "limiting reactant." The reactant that remains is called the "excess reactant."
- By using the ideal gas law, you can use the volumes of gases in stoichiometric calculations.

Chapter 20

Qualitative Chemical Kinetics

In This Chapter

- What kinetics is
- Energy diagrams
- Factors that affect reaction rates

You're already familiar with the idea that some chemical reactions are fast and some are slow. For example, if you place a match to a log in your fireplace, a very slow combustion reaction takes place, causing the log to convert to ashes, water vapor, and carbon dioxide over a period of a few hours. On the other hand, if you place a match to a similar mass of gasoline, you would expect the resulting combustion reaction to take place over a shorter period of time.

In this chapter, we're going to study chemical reaction rates without all of the numbers and fancy equations. Chapter 21 will go into much greater mathematical detail, but it's probably a good idea to start off with a quick review.

What Is Kinetics?

It's often important to know how quickly a chemical reaction will proceed. To do this, we generally study how quickly the reactants in a chemical process vanish and how quickly the products are formed. The study of reaction rates is called *kinetics*.

Kinetics is important because it allows us to do the following very handy things:

- We can tweak the chemical reaction so it proceeds more or less quickly (depending on what we want to do). We could cook a frozen burrito in an oven at 250°, but it would take a very long time. By turning the oven to "broil" (~600°F), frozen burritos cook in less than 10 minutes! Likewise, we can save a lot of money by figuring out a way to keep a steel ship from rusting so quickly.
- Kinetics enables us to figure out how the chemicals in a reaction combine with one another. Let's say I'm performing a chemical reaction, but it's going much too slowly. However, I've got the bright idea to add double the amount of one of the reactants to see if it will speed up the reaction. If the reaction rate doesn't change, that tells me that the reactant I've added doesn't have much to do with the process that determines the rate. If the beaker foams over, I know that the concentration of the reactant has a great deal to do with reaction rate!

def•i•ni•tion

Kinetics is the study of chemical reaction rates.

So, without further ado, let's learn some kinetics!

Energy Diagrams

One of the ways to describe how a chemical reaction occurs is with something called an "energy diagram." An energy diagram is a graph that shows the amount of energy that the reactants have at all points throughout the chemical reaction.

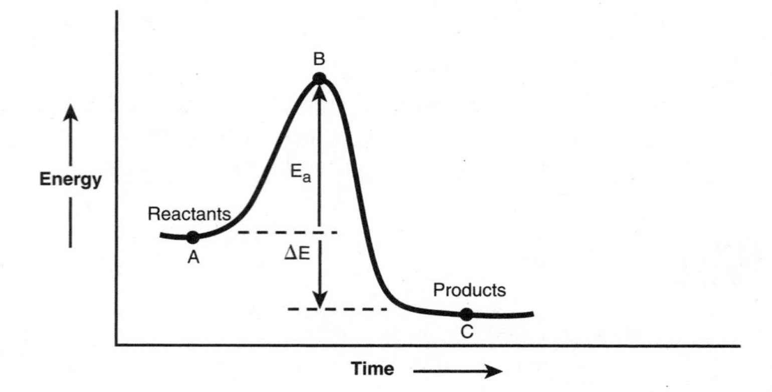

Figure 20.1

A typical energy diagram.

You may notice that this energy diagram looks like a hill—in fact, it behaves very much like one. Let's imagine that we're the reactants in a chemical reaction and are about to go on an adventure that will transform us into the products. We'll start at the left side of the diagram (point A) and walk uphill toward the right.

In this particular reaction, our little feet quickly get tired as we walk up the hill toward point B because climbing uphill requires energy. In fact, you might say that there's a barrier between us and our destination. Our "energy barrier" (the energy difference between point A and point B) is called the *activation energy* (E_a), because we will need at least this much energy to be converted into the products of the chemical reaction.

def•i•ni•tion

The **activation energy** is the minimum quantity of energy that's required for the reactants to form products.

If we don't have enough energy to get to the top of the hill, we'll get too tired to continue and just return to our starting point without changing at all. What we find is that during a chemical reaction, we can't stop right in the middle of a reaction—the reactants either don't do anything or they become products.

However, let's say we have enough energy to climb the hill (point B). Point B is called the "transition state" or "activated complex" of the chemical reaction because that's the point where the reactants are halfway toward turning into the products. This activated complex doesn't look like the reactants of the chemical reaction anymore because the reactants have started to react with each other. However, it doesn't look like the finished products, either. Typically, the activated complex of a chemical reaction lasts for only a very short period of time—it either finishes reacting to become products or goes back down the hill where it came from to regenerate the original reactants.

The Mole Says

To give you a better idea of what a transition state looks like, imagine spilling a cup of coffee in your lap. The "reactants" for this process are the coffee in a cup and your lap. The "product" of this process is your lap covered in hot coffee. The transition state of this process would be the point where you'd spilled the coffee from the cup, but it hadn't yet hit your lap—it looks like neither the reactants nor the product, but something in between.

Once we start moving down the hill, it's smooth sailing to become the products of the reaction. When we hit point C, the reaction has been completed and our destination has been reached.

One other value on this table that's important is the energy difference between point A and point C. To people, this is analogous to the altitude change from our starting and ending points. To a chemical reaction, this value (ΔE) is the difference in energy

You've Got Problems

Problem 1: Sketch the energy diagram for a reaction in which ΔE is positive.

between the products and reactants. If ΔE is positive, the products have more energy than the reactants, so the total energy change for this reaction is positive. If ΔE is negative (as it was in our example), the reactants have more energy than the products, resulting in an overall negative energy change.

Reversibility and Equilibrium

Sometimes, a chemical reaction has a very low activation energy. Such an example is shown in the following figure.

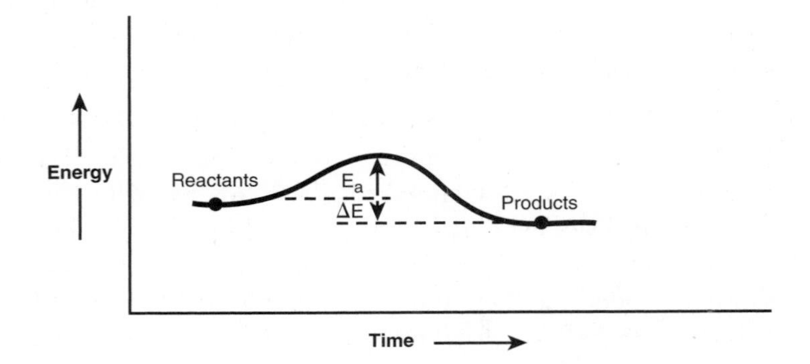

Figure 20.2

The energy diagram for a reversible reaction.

In reversible reactions, the products can frequently react with one another to regenerate the original reactants. To put this in equation form, the process can either go forward ($A + B \rightarrow C$) or backward ($C \leftarrow A + B$). Reversible processes such as this example are called *equilibria* because when the reaction has gone to completion, there are constant amounts of A, B, and C present. We'll discuss equilibria in much greater detail in Chapter 22.

def•i•ni•tion

A reversible reaction is one in which the reactants form products and the products re-form reactants. When the concentrations of the products and reactants have stabilized (the rates of the forward and backward processes are the same), the reaction is said to be at **equilibrium**.

Please note that all chemical reactions are reversible, meaning that for all reactions, the products can react to regenerate the reactants. However, if the activation energy for the reverse process is large enough, this reverse reaction is so slow as to be insignificant when compared to the forward reaction. Because many reactions fall under this category, the single-headed arrow $\rightarrow$ is commonly (but incorrectly) used to describe these reactions.

Factors That Change Reaction Rates

There are a number of different factors that can change the rate of a chemical reaction. In this section, we'll discuss a few of them.

Temperature

The temperature of the reactants affects the rates of chemical reactions. For example, if I were to put a roast in the oven at 150° F, it would take a very, very long time to cook because the process of cooking is very slow at low temperatures.

However, let's imagine that my neighbor came over and cranked up the oven temperature to 500°F. As you might expect, the roast would cook much more quickly—by the time the stove alarm went off, I'd have nothing more than a smoky lump of charcoal.

Why do reactions occur more quickly at high temperatures? As we discussed before, the activation energy of a chemical reaction determines the rate of the reaction. Consequently, if the reactants don't have enough energy to form products, the reaction doesn't go anywhere. To cook a roast, there simply isn't enough energy at 150°F for the molecules in the roast to undergo chemical changes at a useful rate. However, if we increase the energy available to the reactants by increasing the temperature, the molecules in the roast will react at a much faster pace.

> **Chemistrivia**
>
> A good rule of thumb is that chemical reaction rates increase roughly by a factor of 2 for every 10 degrees Celsius of temperature increase.

For those of you who like looking at graphs, we can express this in terms of the fraction of molecules that have enough energy to react.

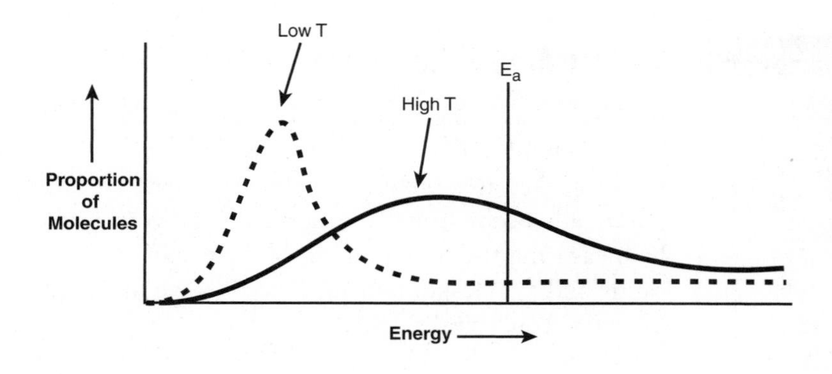

Figure 20.3

The dotted line represents the energies of reactant molecules at low temperature, and the solid line represents the energies of reactant molecules at high temperature.

You've Got Problems

Problem 2: Using kinetics, explain why you need to put ice in a cooler if you're taking egg salad sandwiches on a picnic.

In this diagram, we can see that at low temperatures (the dotted line), very few of the molecules have enough energy to undergo the combustion reaction. However, at high temperature (the solid line), more of the molecules are able to react. In the case of my roast, more of the meat molecules can cook at elevated temperatures because they have a higher average energy.

Concentration of the Reactants

I dislike crowds of people. There's something about a bunch of sweaty people bumping into me that drives me crazy. Every time somebody bumps me in a crowd, my reaction is inevitably to say "Hey, watch it, buddy!"

Let's imagine what would happen if I was walking through an airport that had only five other people in the terminal. Though one or two of them would find a way to bump into me ("Hey, watch it buddy!"), we would probably collide only very occasionally.

On the other hand, if I were to travel during a holiday weekend, I'd find a terminal full of people. As a result, I'd always be bumping into people, causing me to react more and more frequently ("Hey, watch it, buddy!").

The same thing happens when you increase the concentration of the reactants in a chemical reaction. Because there are more reactant molecules present, they're more likely to bump into each other, resulting in a chemical reaction. Of course, not every collision will result in a chemical reaction because not every molecule will have obtained a suitable amount of energy. However, the likelihood of energetic molecules bumping into each other increases, causing the reaction rate to increase.

Chemistrivia

Activated charcoal is used in fish filters to remove organic wastes. Though the pieces of charcoal look reasonably large, they have many small crevices that increase their surface areas. As a result, a small amount of activated charcoal can remove a very large quantity of fish poop.

Surface Area of Reactants

An object reacts more quickly if it's broken into smaller chunks. This is because small objects have a larger surface area on which a chemical reaction can potentially occur. You can observe this on your own by measuring the time it takes for a piece of hard candy and a tablespoon of sugar to dissolve in a glass of water. You'll find that the tablespoon of sugar dissolves far more quickly because the surface area of the small crystals is much larger.

To see that small objects have larger surface areas than large ones, take a look at the accompanying figure, which shows the surface area of an 8 cm^3 cube versus that of 8 1 cm^3 cubes. Both the large cube and the smaller cubes have the same total volume, but they have different surface areas.

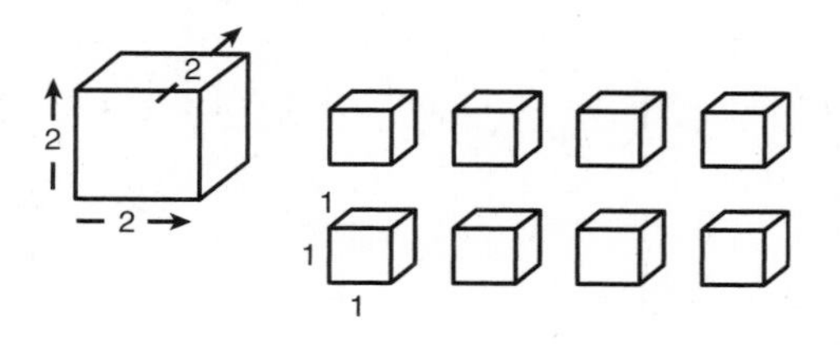

Figure 20.4

The surface area of one 8 cm^3 cube is 24 cm^2, while the surface area of eight 1 cm^3 cubes is 48 cm^2. As a result, the small cubes react more quickly than the large one.

By dissolving a reactant, you can decrease its size down to that of an individual molecule, vastly increasing its surface area. Conceptually, dissolving a material is the same as dividing the cube above into an infinitely smaller number of cubes with a much larger surface area. As a result, chemicals react more quickly as liquids or gases, or in solution, than they do as solids.

You've Got Problems

Problem 3: Using kinetics, explain why you need to grind up coffee beans before putting them in the coffee filter.

Catalysts

Catalysts are materials that increase the rates of chemical reactions without actually being consumed. Common examples of catalysts are platinum (which is used in automobile catalytic converters to convert CO and NO into CO_2 and N_2) and the enzymes in your body that cause biochemical reactions to proceed more quickly.

Catalysts can be compared to my friend Erika (not that you know her). When I was in college, Erika was roommates with a woman named Ingrid. Erika helped to set us up, and I ended up marrying Ingrid. Though Erika was involved in the process (by introducing me to her roommate), she didn't actually take part in the resulting reaction. Therefore, she was a catalyst for that process. I bet you never considered the romantic possibilities of catalysis!

def•i•ni•tion

Catalysts are materials that increase the rates of chemical reactions without actually being consumed. They work by creating new pathways with lower activation energies for forming products.

To see how catalysts work in chemical reactions, let's take a look at the energy diagram for the reaction of A + B → C.

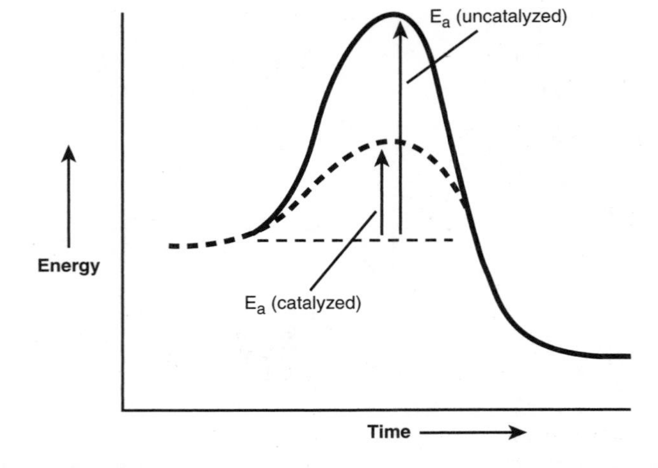

Figure 20.5

Catalysts work by reducing the activation energies of chemical reactions.

Chemistrivia

Inhibitors are chemicals that do the opposite of catalysts—they cause reactions to be slowed or stopped rather than accelerated. When inhibitors are used, the reaction is said to be "poisoned" because it no longer takes place as efficiently.

In this diagram, A and B require a large amount of energy to form compound C. By introducing a catalyst, A and B are pushed together in such a way that less energy is required for the reaction to take place. Because the activation energy for this process is lower, the reaction rate increases. However, from the diagram, note that the energies of the reactants and products remain the same, though a new, lower activation energy pathway is used to get from one to the other.

The Least You Need to Know

- Kinetics is the study of chemical reaction rates.
- Energy diagrams show how the energies of the reactants change during a chemical reaction.
- All reactions are reversible, but in many cases, the backward reaction is so much smaller than the forward reaction as to make it *effectively* irreversible. When the concentrations of product and reactant stay constant, the system is said to be in equilibrium.
- Temperature, reactant concentration, surface area, and catalysts can all change the rates of a chemical reaction.

Chapter 21

Quantitative Chemical Kinetics

In This Chapter

- Differential rate laws
- Integrated rate laws
- Half-lives
- The Arrhenius Equation
- Reaction mechanisms

It's time to talk numbers. In the last chapter, we discussed the qualitative aspects of chemical kinetics. You learned about some of the factors that can affect the rate of a chemical reaction and about energy diagrams.

Now that you understand why things happen in a qualitative sense, it's time to explain chemical rates in numerical terms. It's nice to know that we can make a reaction move more quickly by heating it, but it's even nicer to know how long that process is likely to take. After all, it's great to be able to double a reaction rate, but if that cuts the reaction time from 10 to 5 weeks, we're still going to be sitting around twiddling our fingers and toes while we wait for it to finish.

Fortunately, there are tricks we can use to make this process easier. By the time you finish this chapter, you'll be a kinetics pro!

What Are Rate Laws?

Rate laws are expressions that show how the rate of a chemical reaction depends on the concentration or temperature of the reactants. There are two types of rate laws that are commonly discussed: differential rate laws and integrated rate laws. When reading this chapter, keep in mind that both types of rate laws are derived from the same information, but are written in different forms to solve different types of problems.

The Mole Says

A couple of chapters back, I mentioned that I was going to use the $\rightleftharpoons$ arrows instead of the → arrow in chemical reactions because all reactions can proceed in both the forward and backward directions. In this chapter, we're going to assume that the backward reaction is so slow that it can be ignored, so we'll just use the → arrow to describe chemical reactions.

Differential Rate Laws

There are two main types of rate laws we'll be discussing in this chapter. The first, called a *differential rate law*, describes how the concentration of the reactants affects the reaction rate.

Let's see what a differential rate law looks like, and let's define the terms so we can get started:

For a reaction where A + B + … → C, the rate law is:

$$\text{Rate} = k[A]^x[B]^y \ldots$$

Whoa! Let's explain what the terms in this equation mean. The values "[A]" and "[B]" represent the concentrations of compound A and compound B in moles per liter (M). "k" is the rate constant for this reaction (rate constants indicate the speed at which a reaction takes place and are different for each reaction). "x" and "y" are referred to as the "orders" of each reactant, and their sum (x+y) is the "reaction order." Both "x" and "y" are determined experimentally, and as we'll see later in this chapter, can give us useful tips about how a chemical reaction takes place.

def•i•ni•tion

A **differential rate law** explains the relationship between the concentration of the reactants and the reaction rate.

Let's do an example where we use experimental data to derive the rate law for a chemical reaction.

Example: We've found the initial rates of a chemical reaction (A + B → C) with the following initial concentrations of compound A and B.

Experiment	[A](M)	[B](M)	Initial Rate (M/s)
1	0.0100	0.0100	3.00×10^{-5}
2	0.0200	0.0100	6.00×10^{-5}
3	0.0100	0.0200	6.00×10^{-5}

Using this information, determine the rate law for this reaction, the value of the rate constant, and the initial rate of the reaction when the concentration of compound A is 0.0500 M and the concentration of compound B is 0.0400 M.

The Mole Says

The rate constant (k) for a reaction doesn't change when the concentrations of the reactants are changed. However, changing the concentrations of the reactants will change the rate of a chemical reaction because of the [A] and [B] terms in the rate equation.

Solution:

To find the rate law, we use the equation

$$\text{Rate} = k[A]^x[B]^y.$$

From our initial data, we can see that the rate doubled between experiment 1 and experiment 2. This occurred because we doubled the concentration of compound A. As a result, the reaction rate seems directly proportional to A, making it first-order with respect to A because $x = 1$. Likewise, the reaction rate doubled between experiment 1 and 3 when we doubled the concentration of compound B. As a result, the reaction is first order with respect to B ($y = 1$).

Plugging our experimentally derived values of x and y into the equation, we get a rate law of

$$\text{Rate} = k[A][B].$$

To determine the rate constant, we need to use our new and improved rate law to find k. Let's use our data from experiment 1 to do this

$$3.00 \times 10^{-5}\ \text{M/s} = k(0.0100\ \text{M})(0.0100\ \text{M})$$

$$k = 0.300\ \text{M/s}.$$

To find the rate of this reaction when the concentration of A is 0.0500 M and the concentration of B is 0.0400 M, we need to use the rate law we found earlier (rate = k[A][B]) and the value of k we just found (k = 0.300 M/s). Plugging the values of [A], [B], and k into this equation gives us the following.

Rate = k[A][B]

Rate = (0.300 M/s)(0.0500 M)(0.0400 M)

Rate = 6.00×10^{-4} M/s

It's as simple as that!

You've Got Problems

Problem 1: I've just performed the chemical reaction A + B → C and determined the following initial reaction rates in three experiments under the following conditions.

Experiment	[A]	[B]	Initial Rate (M/s)
1	0.0100	0.0100	5.00×10^{-6}
2	0.0200	0.0100	2.00×10^{-5}
3	0.0100	0.0200	5.00×10^{-6}

Using this information, find the rate law and rate constant for this reaction. Once you have these, determine the rate of the reaction if the initial concentration of A was 0.0150 M and the initial concentration of B was 0.0250 M.

Integrated Rate Laws

The second main type of rate law describes how the concentrations of the reactants change as the reaction progresses. These are commonly referred to as *integrated rate laws.*

To make our lives easier, we'll explain how we can determine the rate laws for various chemical reactions.

def•i•ni•tion

Integrated rate laws describe how the concentrations of the reactants in a chemical reaction vary over time.

First-Order Integrated Rate Laws

Let's say that we're studying the rate of reaction A → B and that we've already determined that this is a first-order reaction.

The rate law for this reaction is:

Rate = k[A].

Chemistrivia

For those of you who are interested in math, the term "ln" stands for "natural logarithm." The term "ln[A]" stands for the power of the mathematical constant "e" that generates [A].

What we need to do is manipulate this equation in such a way that we can relate the concentration of reactant A to the amount of time that the reaction has been proceeding. Some bad news—this requires calculus, which I don't really feel like going into. However, when we do this manipulation, we find that:

$$\ln[A] = -kt + \ln[A_o]$$

where k is the rate constant of this reaction, t is the amount of time since the start of the reaction (in seconds), $[A_o]$ is the initial concentration of reactant A, and [A] is the concentration of reactant A after t seconds have elapsed.

This equation is really neat for the following reasons.

- It allows us to determine how the concentration of our reactant changes as the reaction proceeds (as long as we know what $[A_o]$, t, and k are).
- This equation is in the form $y = mx + b$, which, as you may know, is the equation for a straight line. For this type of equation, when y is plotted against x, the slope of the line is m and the y-intercept is b. Because y in this equation is "ln[A]", m is "–k", and b is "$\ln[A_o]$", we can get some pretty handy information by plotting ln[A] versus the time in seconds—the slope of this line will be the negative of the rate constant (–k) and the y-intercept will be the natural logarithm of the initial concentration of compound A ($\ln[A_o]$).

The Mole Says

In first-order reactions, a plot of ln[A] vs. t will always give you a straight line. The slope of the line is –k and the y-intercept is $\ln[A_o]$. This is useful, because if you're not sure about the order of a reaction, you can figure it out by graphing ln[A] vs. t—if it's a straight line, it's a first-order reaction!

Now that we understand the relationship between the reactant concentration and time for a first-order reaction, we can determine something known as the "half-life" for these reactions.

You've Got Problems

Problem 2: For the reaction A → B, the following kinetic data were collected.

[A]	Time (s)
0.0650	0
0.0460	25
0.0325	50
0.0163	100
0.0081	150

Using these data, prove that this is a first-order reaction and determine the value of k.

What, you might be asking, is a half-life? The half-life of a chemical reaction is the amount of time it takes for half of the reactant to be converted to product. Half-life is denoted by the symbol $t_{1/2}$. The following figure displays the half-lives for a chemical reaction:

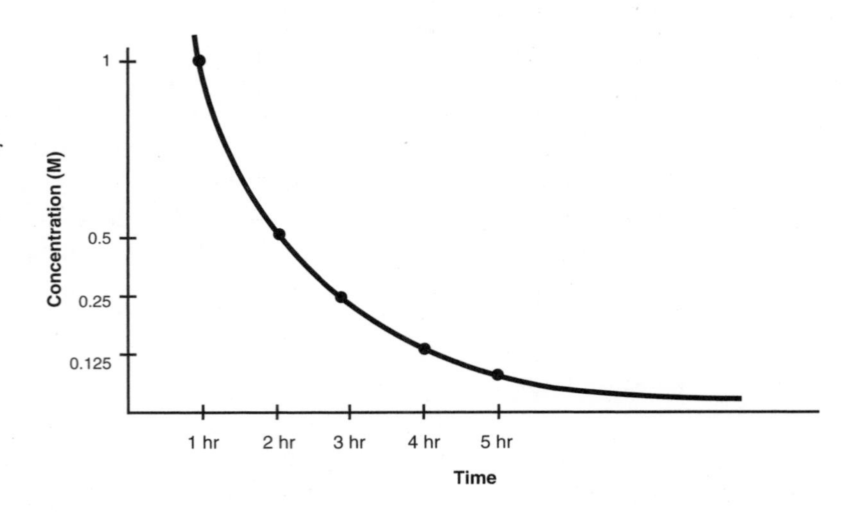

Figure 21.1

A graph showing the half-lives for a chemical reaction. Though the concentration of the reactants decreases, the half-life stays the same over the life of the reaction for first-order processes.

You've Got Problems

Problem 3: Determine the rate constant for a first-order reaction with a half-life of 65 seconds.

To determine how long the half-life of a first-order chemical reaction is, we use the equation we used to determine the relationship between concentration and time:

$$\ln[A] = -kt + \ln[A_o]$$

At $t = t_{1/2}$, the concentration of A is exactly $[A_o]/2$, half of its original value. When we substitute $[A_o]/2$ for [A] in this equation and rearrange, we ultimately end up with:

$$t_{1/2} = \frac{0.693}{k} \quad .$$

Second-Order Integrated Rate Laws

Let's consider the reaction A → B, where the reaction rate is second-order with respect to the concentration of A. The rate law for this reaction is:

Rate = $k[A]^2$.

We can use the miracle of calculus to determine how the concentration of A varies with time, much like we did with first-order reactions to find that:

$$\frac{1}{[A]} = kt + \frac{1}{[A_o]} \quad .$$

$[A_o]$ represents the initial concentration of A, [A] represents the concentration of A at time t, and k represents the rate constant.

This equation describes a straight line with the general form y = mx + b, so if you graph 1/[A] (y-axis) vs. the time (x-axis), the slope of the line is equal to the rate constant, and its y-intercept is $1/[A_o]$.

This leads us to a handy way of solving problems. If we have a process where A → B, we can determine whether it's first-order or second-order in [A] by making two graphs.

- The first graph would plot ln[A] vs. t. If this resulted in a straight line, the process would be first-order in [A].
- The second graph would plot 1/[A] vs. t. If this graph resulted in a straight line, the process would be second-order in [A].

For second-order reactions, the half-life is no longer constant. In other words, if the first half-life of a reaction is 10 minutes, the second half-life will be something different. Because this is more complicated than we really want to worry about here, we're not going to worry about it.

Zero-Order Integrated Rate Laws

Sometimes, the rate of a chemical reaction doesn't depend at all on the concentration of the reagent. This is particularly common for reactions involving catalysts because the quantity of catalyst present is mainly responsible for the rate of reaction. In such a case, the integrated rate law is:

$$\text{Rate} = k[A]^0 = k.$$

Simply put, the rate is equal to the rate constant because anything raised to the zeroth power is 1.

The Dependence of Reaction Rate on Temperature

In Chapter 20, we discussed how increased temperatures cause the rate of a chemical reaction to increase. We explained this by saying that as reactant temperature increases, more of the molecules have the required activation energy for the reaction to take place.

In 1889, the Swedish chemist Svante Arrhenius was studying the dependence of reaction rate on temperature and discovered that most reaction rate data fit the following equation.

$$k = Ae^{-E_a/RT}$$

In the Arrhenius equation, "k" is the rate constant for the reaction, E_a is the activation energy, R is the ideal gas constant, T is the temperature (in Kelvins), and A is a constant called the "frequency factor." The frequency factor in the Arrhenius equation reflects the probability that the molecules colliding will be in a proper orientation to undergo a chemical reaction.

To simplify this equation, we can take the natural logarithm (the "ln" button on your calculator) of this equation to find that:

$$\ln(k) = \left(-\frac{E_a}{R}\right)\left(\frac{1}{T}\right) + \ln(A)\ .$$

Again, we have an equation with the form y = mx + b, so a graph of ln(k)vs. 1/T will yield a straight line with a slope equal to $-E_a/R$ and a y-intercept equal to ln(a). By making a simple (well, maybe not so simple) graph, the activation energy of the reaction can be determined.

However, there's an easier way to find the activation energy of a reaction. Let's assume that we can find the rate constants of a reaction at two different temperatures. Using the Arrhenius equation and a bunch of fancy math (that we'll conveniently skip over), we find that:

$$\ln\left(\frac{k_2}{k_1}\right) = \left(\frac{E_a}{R}\right)\left(\frac{1}{T_1} - \frac{1}{T_2}\right).$$

In this equation, T_1 is the first temperature and k_1 is the rate constant at this temperature. T_2 is the second temperature, with k_2 the rate constant at this temperature. By putting experimental data into this equation, we can determine the activation energy of a reaction.

The Mole Says

The activation energy of a reaction doesn't depend on the temperature—it always has the same value. The reaction rate increases as we raise the temperature because more molecules have the minimum activation energy required for the reaction to proceed.

Example: The reaction for A + B → C has a rate constant of 1.20 L/mol·s at 140° C and a rate constant of 2.50×10^5 L/mol·s at 340° C. Given that R = 8.31 J/K·mol, what is the activation energy for this reaction?

Solution: T_1 = 413 K (140° C + 273), k_1 = 1.20 L/mol·s, T_2 = 613 K (340° C + 273), $k_2 = 2.50 \times 10^5$ L/mol·s. Substituting these values and the value of R into the preceding equation, we find that:

$$\ln\left(\frac{k_2}{k_1}\right) = \left(\frac{E_a}{R}\right)\left(\frac{1}{T_1} - \frac{1}{T_2}\right)$$

$$\ln\left(\frac{2.50x10^5 L/s}{1.20 L/s}\right) = \left(\frac{E_a}{8.31 J/K \cdot mol}\right)\left(\frac{1}{413K} - \frac{1}{613K}\right)$$

$$\ln(2.08x10^5) = \left(\frac{E_a}{8.31 J/K \cdot mol}\right) 0.000790 K^{-1}$$

$E_a = 1.29 \times 10^5$ J/mol

You've Got Problems

Problem 4: For the reaction A + B → C, the rate constant at 800° C is 2.50 L/mol·s and the rate constant at 850° C is 122 L/mol·s. Given that R is 8.31 J/K·mol, what is the activation energy for this reaction?

Reaction Mechanisms

In what order do the reactants combine with one another in a chemical reaction? Read on to find out!

An Introduction to Mechanisms

Now that we've discussed rate laws in great gory detail, it's time to learn how we can use them to figure out the way reactants in a chemical process form products. The method by which a chemical reaction takes place is known as the *reaction mechanism*.

Let's take a look at a hypothetical reaction mechanism for making scrambled eggs.

Step 1: Raw eggs in shell → Raw egg goo + yucky shells

Step 2: Raw egg goo + milk + butter → scrambled eggs

There are some important things we need to understand before we can continue talking about reaction mechanisms.

- Each of the steps in a chemical reaction is called an elementary reaction (also called an elementary step).
- When you add all of the elementary steps in a chemical reaction together, you should end up with the final chemical reaction. In our example, adding the two elementary steps gives us:

  Raw eggs in shell + raw egg goo + milk + butter → raw egg goo + yucky shells + scrambled eggs.

  After canceling "raw egg goo" from both sides of the equation, we end up with the overall equation:

  Raw eggs in shell + milk + butter → scrambled eggs + yucky shells (which we'll just throw away).
- Chemicals formed by one step and consumed by another are called "intermediates" because they are merely stepping stones to the final products. To make scrambled eggs for breakfast, we need to break the eggs to get them out of the shells. The raw egg goo in our example is an intermediate because we didn't start with raw egg goo (we started with regular eggs) and we won't end up with raw eggs because we'll cook them to make scrambled eggs. They exist only for a short time during the process.
- The "molecularity" of a step is defined as the number of reactant molecules that combine in the step. If only one molecule reacts, the step is "unimolecular." If two react, the step is "bimolecular."

The following table indicates the general forms for the rate laws of various elementary reactions, where Z represents the product or products of the reaction.

Molecularity	Step	Rate Law
Unimolecular	$A \rightarrow Z$	Rate = $k[A]$
Bimolecular	$2A \rightarrow Z$	Rate = $k[A]^2$
Bimolecular	$A + B \rightarrow Z$	Rate = $k[A][B]$

Molecularity	Step	Rate Law
Termolecular	$3A \rightarrow Z$	Rate = $k[A]^3$
Termolecular	$2A + B \rightarrow Z$	Rate = $k[A]^2[B]$
Termolecular	$A + B + C \rightarrow Z$	Rate = $k[A][B][C]$

Now, you might expect the overall rate laws for a multi-step process to be difficult to understand. After all, if we have two elementary reactions in a mechanism, don't we have two rate laws to keep track of at once?

Not necessarily. Let's go back to the scrambled egg-making example:

Step 1: Raw eggs in shell → Raw egg goo + yucky shells

Step 2: Raw egg goo + milk + butter → scrambled eggs

Which process is faster? If you've ever made scrambled eggs, you know that cracking an egg is a very fast process, while it takes a few minutes to actually cook it into something edible. As a result, the speed at which we crack the eggs has very little to do with how fast the eggs are made—it's the much slower rate at which we can cook the eggs that determines the speed of this process.

In a reaction with several elementary steps, the slowest step is called the "rate-determining step" because it alone is responsible for the observed overall rate of the chemical reaction.

Proving a Reaction Mechanism Invalid

Getting back to the main point of this topic, how can we use rate laws to determine the mechanism for a chemical reaction? Let's consider the following process.

$$A + B \rightarrow Y + Z$$

Now, nobody's sure how this process actually occurs, but it is known that the overall reaction rate can be expressed by the following equation.

$$\text{Rate} = k[B]$$

I believe this reaction proceeds via the following mechanism:

Step 1: $2\,A \rightarrow Y + D$ Slow Process

Step 2: $B + D \rightarrow A + Z$ Fast Process

How do I prove this to be either valid or invalid?

Reaction mechanisms are only supported if they conform to the following two rules:

1. The overall equation for the reaction needs to equal the sum of the elementary steps.
2. The rate law for the rate-determining step should match the rate law for the overall reaction.

Let's see if my proposed mechanism works. For the first rule, the elementary steps, when added, should yield the overall equation for the reaction. In my case, the elementary steps add up in the following way:

$$2\,A \rightarrow Y + D$$

$$\underline{B + D \rightarrow A + Z}$$

$$2\,A + B + D \rightarrow Y + D + A + Z$$

which, after canceling out the terms that are present on both sides of the equation, equals the overall equation for this reaction: $A + B \rightarrow Y + Z$. So far, my mechanism is looking pretty good!

Now for the second rule: Does the rate-determining step in my mechanism match the rate law for the overall reaction? In my mechanism, I defined the rate determining step as:

$$2\,A \rightarrow Y + D$$

As a result, we'd expect the rate of the reaction to follow the equation:

$$\text{rate} = k[A]^2$$

This doesn't match the known rate of the reaction, which is equal to k[B]. As a result, my mechanism is invalid.

The Least You Need to Know

- Rate laws are expressions that show how the rate of a chemical reaction depends on the concentrations of the reactants or the time elapsed since the reaction began.
- The half-life of a reaction is the period of time required for half of the reactants to be converted into products.
- The Arrhenius equation is used to determine the relationship between rate and temperature.
- Reaction mechanisms describe the process by which the reactants in a chemical process are converted to products.

Chapter 22

Solution Chemistry/Chemical Equilibria

In This Chapter

- Equilibrium constants
- Heterogeneous equilibria
- Reaction quotients
- Le Châtelier's Principle

In Chapter 20, I briefly mentioned that chemical reactions can go forward or backward, meaning that reactants form products and products re-create reactants. However, I then discussed chemical reactions as if they went only in the forward direction. I could get away with this because many chemical reactions go mostly forward, with only a smidgen of backward reaction.

Well, boys and girls, it's time to stop ignoring the backward reaction! In this chapter, we're going to learn all about chemical equilibria. Hold on to your hats!

What Is an Equilibrium?

Back in Chapter 20 when we discussed energy diagrams, we mentioned that chemical reactions can move in the backward direction as well as the forward direction, forming reactants from the products. Such a reaction is referred to as being "reversible," and has the equation $A \rightleftharpoons B$.

In Chapter 21, we learned that as the concentration of reactants increases, the rate of the reaction increases. As you might imagine, the same is true for the products—the reverse reaction will tend to speed up as the concentrations of the products increases. As a result, the concentrations of the reactants and products in a reversible reaction change in the following way over time.

- At the beginning of a reaction, the reactants are the only chemicals present in the beaker. As a result, the forward reaction is the only one that takes place.
- Shortly after the start of the reaction, the concentra tants is decreasing because they are being consumed to make products—as a result, the forward reaction is decelerating. At the same time, the concentration of the products is small but increasing, causing the reverse reaction to accelerate.
- At some point in the reaction, the rates of the forward and reverse reactions will become equal. When this occurs, the concentrations of the products and reactants cease changing and the system is said to be at *equilibrium*.

def•i•ni•tion

When the forward and reverse reactions for a reversible reaction take place at the same rate, the system is at **equilibrium** and the concentrations of both the products and reactants stay the same.

A lot of new chemists think that when a reaction has reached equilibrium, the process completely stops. Pardon my French, but *non!* Both the forward and reverse reactions are still taking place—they just take place at the same rate. As a result, the reaction may appear to the casual observer to have stopped even though molecules are still making the conversion between products and reactants.

Equilibrium Constants

All equilibria have different ratios of products to reactants. As an example, let's consider a reaction in which the forward reaction takes place quickly and the reverse reaction is slow. When the system reaches equilibrium, a vast majority of the chemicals in the beaker will be products (we usually say that the "equilibrium favors the products"). On the other hand, if the forward reaction is slow and the reverse reaction is fast, the equilibrium favors the reactants.

It probably won't be a big surprise to you to find that scientists have figured out a way to quantify the position of an equilibrium. This was done by the "law of mass action."

What the law of mass action says is simple. Let's say that we have a reaction taking place in solution with the following equation.

$$aA + bB \rightleftharpoons cC + dD$$

The equilibrium condition can be expressed using the following equation:

$$K_{eq} = \frac{[C]^c[D]^d}{[A]^a[B]^b}$$

where K_{eq} is a constant for the process called the "equilibrium constant," each of the letters in brackets stands for the concentration of that chemical in mol/L (M) when the reaction has reached equilibrium, and the superscripts stand for the coefficient of each chemical. For gases, the equilibrium constant is determined in almost the same way, except that partial pressures are used in place of concentrations.

The Mole Says

We can relate the equilibrium constant of a chemical equilibrium to the rates of the forward and reverse reactions. For example, consider the process A   B. The rate of the forward reaction, $A \rightarrow B$, is $k_f[A]$, where the little "f" after the k denotes the rate constant for the forward reaction. Likewise, the rate of the reverse reaction, $B \rightarrow A$, is $k_r[B]$, where "r" denotes the rate constant for the reverse process. Since at equilibrium, the rates for the forward and backward reactions are the same ($k_f[A] = k_r[B]$), we can say that the equilibrium constant, K_{eq}, is equal to [B]/[A] and k_f/k_r.

The equilibrium constant is important because it gives us an idea of where the equilibrium lies. The larger the equilibrium constant, the further the equilibrium lies toward the products. For example, an equilibrium constant of 1.0×10^{-6} suggests that very little of the reactants have formed products at equilibrium, while an equilibrium constant of 1.0×10^{6} suggests that at equilibrium, most of the chemical species present are products.

K_{eq} is a neat constant because it allows us to determine the ratios of the concentrations of each chemical species in the equilibrium, which is pretty handy for reasons we'll see in this example.

Bad Reactions

Be careful not to mistake the equilibrium constant K_{eq} for the rate constant k!

Example: Acetic acid breaks up in water via the following process.

$$C_2H_3O_2H_{(aq)} \rightleftharpoons H^+_{(aq)} + C_2H_3O_2^-{}_{(aq)}$$

Write the expression for the equilibrium constant, and determine what the value of K_{eq} is given these concentrations of each of the chemical species at equilibrium.

Chemical Species	Concentration at Equilibrium (M)
$C_2H_3O_2H$	.68
H^+	3.5×10^{-3}
$C_2H_3O_2^-$	3.5×10^{-3}

To figure out the expression for the chemical equilibrium, use the equation we learned for finding K_c.

$$K_{eq} = \frac{[C_2H_3O_2^-][H^+]}{[C_2H_3O_2H]}$$

To determine the value of K_{eq}, all we need to do is substitute the values given to us for the concentrations in the equilibrium expression. As a result, we get the following.

$$K_{eq} = \frac{(3.5x10^{-3}\text{M})(3.5x10^{-3}\text{M})}{0.68\,\text{M}}$$

$$K_{eq} = 1.8 \times 10^{-5}$$

Once we have an equilibrium constant, we can use it to figure out what the equilibrium concentrations of the products will be given an initial concentration of the reactants. Let's see another example.

Example: Given the reaction $A \rightleftharpoons B + C$, what will be the equilibrium concentrations of B and C if the initial concentration of A is 1.00 M and the equilibrium constant is 1.86×10^{-6}?

Solution: Let's walk through this problem step by step. The first step is to write the equilibrium expression for this process.

$$K_{eq} = \frac{[B][C]}{[A]}$$

Our next step is to figure out what the concentrations of each species will be at equilibrium. We do this by setting up a chart that shows the initial concentrations of all species, how the concentrations of each will change, and what the final concentrations of each species will be. For this process, the chart is given next (don't panic, we'll explain how we got all these values in a minute).

Species	Initial Concentration	Change	Final Concentration
A	1.00 M	–x	(1.00 – x) M
B	0 M	x	x M
C	0 M	x	x M

Okay, let's talk about where these came from.

- The initial concentration of A is defined by the problem as 1.00 M. Between the time that the reaction starts and the system reaches equilibrium, some of it will turn into the products. How much? We have no idea, so we'll just say that the change was "–x." As a result, our final concentration will be the initial concentration minus the amount of change, or (1.00 – x) M.
- The initial concentrations of both B and C are zero because neither was initially present. However, from the equation, you can see that every time a molecule of A breaks apart, one molecule each of B and C are formed. As a result, if the concentration of A decreases by x, the concentration of both products must increase by x. At equilibrium, the concentration of both B and C is x.

To figure out what the concentrations of each species are, we plug these values into the expression for finding the equilibrium constant. Since we know that the equilibrium constant is 1.86×10^{-6}, all we have to do is solve for x in the following expression.

$$1.86x10^{-6} = \frac{[x][x]}{[1.00 - x]}$$

Now, solving for x won't be a lot of fun because we'll need to use the quadratic equation. Let's face it, we don't really want to memorize the quadratic equation, much less use it.

Fortunately, there's a shortcut we can use to get around using the quadratic equation when the K_{eq} values are very small. If K_{eq} is small, then we can safely guess that the amount of product formed (x) will be very, very small when compared to the initial quantity of the reactant. This is because a very small K_{eq} value means that very little product has been formed. As a result, we can omit the "x" in the denominator to simplify the [1.00 – x] term because [1.00 – x] will be roughly equal to 1.00. Our new (and easier) expression to solve is transformed into:

$$1.86x10^{-6} = \frac{[x][x]}{[1.00]}$$

$x = 1.36 \times 10^{-3}$ M

Let's see if this was a good assumption. If the amount of product formed was 1.36×10^{-3} M, then the final concentration of acetic acid will be (1.00 – 0.00136) M, which, within the constraints of significant figure rules, is still 1.00 M. As a result, our assumption was valid and we saved ourselves a lot of mathematical heartache and toil. I love it when I save myself toil!

Our final concentrations of B and C are 1.36×10^{-3} M.

You've Got Problems

Problem 1: Given the reaction $H_2 + I_2 \rightleftharpoons 2\ HI$, find the following:

a) The general expression for the equilibrium constant.

b) The equilibrium concentration of HI if I start with 2.00 M H_2 and 2.00 M I_2 and the equilibrium constant is 5.00.

Heterogeneous Equilibria

All of the equilibria we've been talking about so far have chemical species in the same phase. For solutions, they're all dissolved, and for gaseous equilibria, they're all gases. Equilibria in which all species are in the same phase are called *homogeneous equilibria*.

However, we can also talk about equilibria in which not all of the species are in the same phase. These equilibria are referred to as *heterogeneous equilibria*. An example of a heterogeneous equilibrium would be when an ionic compound partially dissolves in water.

def•i•ni•tion

A **homogeneous equilibrium** occurs when all reagents and products are found in the same phase (solid, liquid, or gas) and a **heterogeneous equilibrium** is when they are in different phases.

The Mole Says

In our example, the equilibrium constant is called "K_{sp}" for "solubility product constant." Any time you study how one compound dissolves in another, you'll need a K_{sp} value.

To demonstrate a heterogeneous equilibrium, we'll discuss the equilibrium expression for when calcium carbonate dissolves in water (a process known as "dissociation"). The equation for this process is:

$$CaCO_{3(s)} \rightleftharpoons Ca^{2+}_{(aq)} + CO_{3\ (aq)}^{2-}.$$

Let's write the equilibrium expression for this process.

$$K_{sp} = \frac{[Ca^{2+}][CO_3^{\ 2-}]}{[CaCO_3]}$$

However, there's a twist. Recall that any time we put something in the square brackets, this means that we need its concentration. Because $CaCO_3$ is a solid in this process, it doesn't really have a clearly defined concentration. As a result, we just leave it out of this expression. (I bet you wish you could do that with everything that didn't make sense!) Likewise, whenever you have a pure solid or a pure liquid (but not a solution) in an equilibrium expression, you leave it out of the expression for K_{eq} or K_{sp}.

Leaving the $[CaCO_3]$ term out of the K_{sp} expression leaves us with the following expression.

$$K_{sp} = [Ca^{+2}][CO_3^{-2}]$$

The Mole Says

Anytime you have a pure solid present in an equilibrium, leave it out of the equilibrium expression entirely.

The K_{sp} value for an ionic compound describes the degree to which the ions are present in a saturated aqueous solution. As with other equilibria, in saturated solutions the salt will dissolve and precipitate out at the same rates, causing no net change in ionic concentrations.

Example: What's the concentration of Ca^{+2} ions in a saturated solution of $CaCO_3$? $K_{sp}(CaCO_3) = 4.5 \times 10^{-9}$.

Solution: In the equation for the dissociation of $CaCO_3$, we see that the concentration of Ca^{+2} and CO_3^{2-} must be the same in a saturated solution because one of each ion is formed by the breakup of each $CaCO_3$. Because we don't know what this concentration will be, let's call it "x."

Putting this into our K_{sp} expression for $CaCO_3$, we find that:

$$K_{sp} = 4.5 \times 10^{-9} = x^2$$

$$x = 6.7 \times 10^{-5} \text{ M.}$$

The concentration of both the calcium and carbonate ions is 6.7×10^{-5} in a saturated $CaCO_3$ solution.

You've Got Problems

Problem 2: What is the equilibrium concentration of bromide ion when enough $PbBr_2$ dissociates in water to make a saturated solution? $K_{sp}(PbBr_2) = 2.1 \times 10^{-6}$.

Disturbing Equilibria: Le Châtelier's Principle

Back in the late 1800s, the French chemist Henri Le Châtelier came up with a rule about equilibria that we still use extensively today, to the chagrin of many chemistry students.

Le Châtelier's Principle states that if you change the conditions of an equilibrium, the equilibrium will shift in a way that minimizes the effects of whatever it is you did.

The Mole Says

Le Châtelier's Principle states that the equilibrium will shift to minimize the effects of whatever you did. However, it's important to note that the concentrations of the chemical species will be different in the new equilibrium than they were before. In essence, a new equilibrium will be created in order to maintain K_{eq} for the reaction.

In other words, equilibria are like obnoxious little kids. For example, if you yell at a little kid, the kid will change his behavior to minimize your yelling. Likewise, if you change the conditions of an equilibrium, it will change in a way that partially undoes whatever it is you did to it in the first place.

Change in Concentration

Let's say that you're doing a reaction with the equation $A + B \rightleftharpoons C$. Le Châtelier's principle says that if we change the concentration, the position of the equilibrium will also change.

For example, if the process explained earlier is at equilibrium, we can disturb the equilibrium by adding a bunch of compound A to the reaction. To minimize the effects of the added compound, the equilibrium will shift in a way that will decrease the amount of compound A—namely, it will produce more of compound C. Likewise, by adding more of compound C, the equilibrium will be pushed toward the left, making more of A and B.

This phenomenon is often seen when an ionic compound dissolves. Let's see an example.

$$CaSO_{4(s)} \rightleftharpoons Ca^{+2}_{(aq)} + SO_4^{-2}{}_{(aq)}$$

The K_{sp} value for calcium sulfate is 2.4×10^{-5}. Doing the math we learned earlier in this chapter, we can easily find the equilibrium concentration of the calcium ion.

$$K_{sp} = 2.4 \times 10^{-5} = [x][x]$$

$$x = 4.9 \times 10^{-3} \text{ M}$$

However, what would happen if we added 1.0 M Li_2SO_4? Because the concentration of the sulfate ion would be increased by 1.0 M, the quantity of the calcium ion would now be.

$$K_{sp} = 2.4 \times 10^{-5} = [x \text{ M}][x + 1.0 \text{ M}]$$

where the concentration of the sulfate ion would be (x + 0.10 M) to compensate for the added sulfate ion. Because the quantity of sulfate ion likely to dissolve is very small compared with the quantity we've added, we'll eliminate the "x" from this term to give us.

$2.4 \times 10^{-5} = (x\ M)(1.0\ M)$

$x = 2.4 \times 10^{-5}\ M.$

As you can see, the quantity of calcium ion has been greatly reduced by the addition of 1.0 M sulfate ion.

The reduced solubility of one compound caused by adding an additional quantity of one of its ions to solution is referred to as the *common ion effect.*

def•i•ni•tion

The **common ion effect** is when the addition of an ion affects the solubility or reactivity of a chemical compound.

Change in Pressure

Gaseous equilibria can be changed by altering the pressure of the gases. For example, let's say that we're doing the reaction $A_{(g)} + B_{(g)} \rightleftharpoons C_{(g)}$. In this reaction, two moles of gas are combining to make one mole of gas.

If we increase the pressure of this mixture of gases by squishing it into a smaller area, the pressure of each of the gases will increase. Le Châtelier's principle states that equilibria tend to want to decrease the effects of any changes, so the equilibrium will shift in a way that reduces the pressure of the system. The only way to accomplish this is to have fewer moles of gas present; as a result, the reaction will shift toward products to reduce the overall pressure.

Bad Reactions

There's another way to increase the pressure of a gaseous mixture: Add another gas that doesn't interact with any of the gases that are in equilibrium (for our example, imagine adding some of gas D to the mixture). It's important to keep in mind that although the total pressure inside the container will increase, the partial pressures of each gas will not. As a result, the addition of gas D won't change the position of the equilibrium at all.

Change in Temperature

Some reactions naturally give off energy (they're exothermic) and some reactions require energy to take place (they're endothermic). If we think of the energy in an exothermic reaction as being a product, it will have the form.

$$A \rightleftharpoons B + \text{energy}.$$

Because endothermic reactions require energy to take place, we can think of energy as a reagent.

$$A + \text{energy} \rightleftharpoons B.$$

When we increase the temperature of a chemical reaction, what we're really doing is adding energy to it. Because we can think of energy as being either a reactant or product, the addition of extra energy will disrupt the equilibrium. For exothermic reactions, the addition of energy will push the reaction to the left toward reactants. For endothermic reactions, the addition of energy will push the reaction to the right toward the products.

You've Got Problems

Problem 3: Determine how the following changes will affect the following equilibrium: $A_{(g)} + 2\ B_{(g)} \rightarrow 2\ C_{(g)} + \text{energy}$

(a) Some of product C is removed from the mixture.

(b) 2 atm of compound D (also a gas) is added to the mixture.

(c) The mixture is squished into a much smaller container.

(d) The temperature of the mixture is decreased.

The Least You Need to Know

- When a system is at equilibrium, the forward and backward reactions occur at the same rate.
- The equilibrium constant describes the position of the equilibrium. Some types of equilibrium constants include K_{eq} for equilibria involving solutions, K_p for equilibria involving gases, and K_{sp} for equilibria involving the dissolution of ionic compounds in water.
- Any pure solids are left out of the equilibrium expression for heterogeneous equilibria.
- Le Châtelier's Principle says that if we alter the conditions of an equilibrium, the equilibrium will shift to minimize the effects of whatever we did.

Chemical Reactions

We've finally gotten to the good part of chemistry! Instead of contenting ourselves with talking about equations and phases of matter, we're at the part where we get to start fires and blow stuff up. At least, that's what always seems to happen when I perform chemical reactions.

If you're interested in seeing the magic of acids and bases, electrochemistry, nuclear reactions, and organic chemistry, keep reading!

Chapter 23

Acids and Bases

In This Chapter

- The three definitions of acids and bases
- The pH scale
- Titrations
- Buffers

As you learned way back in Chapter 18, acid-base reactions have equations with the general form $HA + BOH \rightleftharpoons BA + H_2O$. So far, that's all I've really said about them. As you've probably guessed, there's a lot more to acids and bases than this equation.

The reason that acid-base reactions are so important is that many of the things you come into contact with on a daily basis are either acids or bases. Most fruits are acids, as are carbonated beverages, tea, and battery acid. Common household bases include baking soda, ammonia, soap, and antacids. As you'll find, acids and bases really aren't that difficult to understand once you get the hang of them.

What Are Acids and Bases?

Although I've told you that acids and bases aren't hard to understand, I've got bad news: There are not one but three common definitions used to describe acids and bases: Arrhenius acids and bases, Brønsted-Lowry acids and bases, and Lewis acids and bases. Though this makes it sound as if you'll have to learn about acids and bases three times, the good news is that for many practical purposes, these three definitions are roughly equivalent.

Arrhenius Acids and Bases

Way back in the late 1800s, our old friend Svante Arrhenius from Chapter 21 came up with definitions of acids and bases while working on kinetics problems.

According to Arrhenius, acids are compounds that break up in water to give off hydronium (H^+) ions. A common example of an Arrhenius acid is hydrochloric acid (HCl).

$$HCl \rightleftharpoons H^+ + Cl^-$$

The formulas for acids usually start with hydrogen, though organic acids are a notable exception (more about those in Chapter 25). The names and formulas of some common acids are given in the table below.

Acid Name	Formula
hydrochloric acid	HCl
nitric acid	HNO_3
phosphoric acid	H_3PO_4
sulfuric acid	H_2SO_4
acetic acid	$C_2H_4O_2$ (sometimes written CH_3COOH)

The Mole Says

There are many different names and formulas used to describe the hydronium ion. Though the formula was shown previously as "H^+", it is sometimes written as "H_3O^+" because this is the ion formed when H^+ combines with water. Another common way of referring to hydronium ions is just to call them "protons." This name comes from the fact that H^+ represents a hydrogen atom (one proton and one electron) that has lost its electron, leaving only the bare proton behind.

Arrhenius bases are defined as compounds that cause the formation of the hydroxide ion when placed in water. One example of an Arrhenius base is sodium hydroxide (NaOH).

$$NaOH \rightleftharpoons Na^+ + OH^-$$

Bases typically have "OH" in their formulas, though there are exceptions (most notably with organic acids, *see* Chapter 25). For example, ammonia (NH_3) doesn't contain hydroxide ions but forms them when it reacts with water:

$$NH_3 + H_2O \rightleftharpoons NH_4^+ + OH^-$$

The names and formulas of some common bases are in the following table.

Base Name	**Formula**
ammonia	NH_3
potassium hydroxide	KOH
sodium bicarbonate	$NaHCO_3$
sodium carbonate	Na_2CO_3
sodium hydroxide	NaOH

Some oxides form acids or bases when water is added. Because these compounds don't contain any H^+ or OH^- ions unless they react with water, they're called "anhydrides." Typically, oxides of nonmetals are acid anhydrides (they form acid when placed in water), and oxides of metals are base anhydrides (forming a base when placed in water).

Brønsted-Lowry Acids and Bases

In the early 1900s, an alternate definition for acids and bases was proposed by Johannes Brønsted and Thomas Lowry to account for the fact that ammonia can neutralize the acidity of HCl even if water isn't present. This phenomenon showed them that ammonia is a base, even when there isn't water around to form hydroxide ions.

A Brønsted-Lowry acid is defined as a compound that gives hydronium ions to another compound—for example, hydrochloric acid gives H^+ ions to compounds it reacts with. Brønsted-Lowry bases are compounds that can accept hydronium ions—when ammonia gets a hydronium ion from HCl, it forms the ammonium ion.

The following equation represents the reaction of a Brønsted-Lowry acid with a Brønsted-Lowry base.

$$HNO_3 + NH_3 \rightleftharpoons NO_3^- + NH_4^+$$

In this reaction, nitric acid behaves as an acid because it gives a proton to ammonia. Ammonia behaves as a base because it accepts the proton from nitric acid.

However, if you take a look at the other side of the equation, we find the nitrate and ammonium ions. Because the nitrate ion can accept protons from the ammonium ion (to form HNO_3), the nitrate ion is a very weak Brønsted-Lowry base. Because the ammonium ion has an extra proton to donate (in this case to the nitrate ion), it is a Brønsted-Lowry acid.

> **You've Got Problems**
>
> Problem 1: Identify the conjugate acid-base pairs in the following equation: $H_2SO_4 + HPO_4^{2-} \rightleftharpoons H_2PO_4^{-1} + HSO_4^{-1}$

The nitrate ion is based on the nitric acid molecule, so we say that it is the *conjugate base* of nitric acid. Likewise, the ammonium ion is the *conjugate acid* of ammonia. Together, an acid with its conjugate base (such as HNO_3 and NO_3^-) or a base with its conjugate acid (such as NH_3 and NH_4^+) is referred to as a conjugate acid-base pair.

Lewis Acids and Bases

In the Brønsted-Lowry definition of acids and bases, a base is defined as a compound that can accept a proton. However, *how* does it accept the proton?

One feature that Brønsted-Lowry bases have in common with each other is that they have an unshared pair of electrons. When a hydronium ion comes wandering by the molecule, sometimes the lone pairs will reach out and grab it. An example of this is when ammonia accepts a proton in an acidic solution.

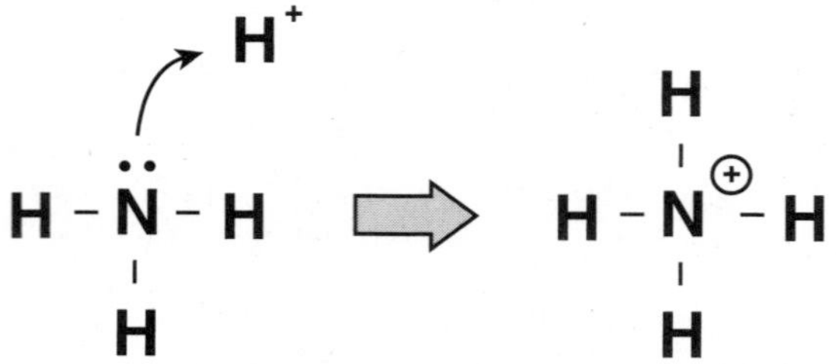

Figure 23.1

Ammonia can grab a proton from nitric acid with its lone pair of electrons.

One way of looking at this process is that the ammonia atom is donating its lone pair to the proton. Because the lone pairs are driving this chemical reaction, we have a new definition of acidity and basicity, called "Lewis acidity/basicity." A *Lewis base* is a compound that donates an electron pair to another compound (the ammonia in our example). A *Lewis acid* is a compound that accepts an electron pair (the H^+ ion in our example).

Though we had ammonia donating a lone pair to a proton in our example, the lone pair in ammonia can react with a lot of other compounds as well. For example, ammonia can donate its lone pair of electrons to BH_3 by the following process.

def•i•ni•tion

Lewis bases are chemicals that can donate electron pairs. **Lewis acids** are chemicals that can accept them.

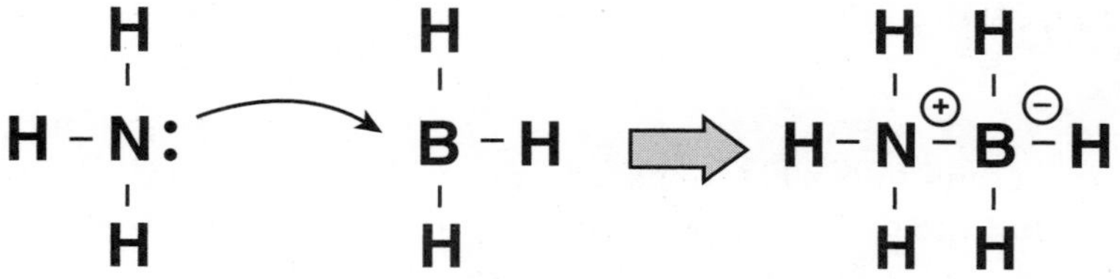

Figure 23.2
The lone pair on ammonia attaching itself to BH_3.

In this process, ammonia is the Lewis base and BH_3 is the Lewis acid.

Generally, the Lewis definition of acids and bases is the most useful because it is the most inclusive of the three definitions. For example, the Brønsted-Lowry definition of an acid includes HF but not BH_3, which doesn't lose a proton when attached by the lone pairs on a Lewis base.

def•i•ni•tion

Arrhenius acids are compounds that form hydronium (H^+) ions in water, and **Arrhenius bases** are compounds that form the hydroxide ion in water.

Brønsted-Lowry acids are compounds that give H^+ ions to other compounds. **Brønsted-Lowry bases** accept H^+ ions.

Lewis bases are chemicals that can donate electron pairs. **Lewis acids** are chemicals that can accept them.

Properties of Acids and Bases

It's frequently possible to tell acids and bases apart from one another by some of their easily observed chemical and physical properties. A table of these properties is shown here.

Property	Acid	Base
Taste	Sour (vinegar)	Bitter (baking soda)
Smell	Frequently burns nose	Usually no smell (except NH_3!)
Texture	Sticky	Slippery
Reactivity	Frequently react with metals to form H_2	React with many oils and fats

The pH Scale

As you might imagine, it's useful to be able to measure the acidity of solutions. For example, the shaving cream I use in the morning has an acid listed on the ingredient label. Clearly, I would have a less enjoyable shaving experience if my shaving cream were as acidic as battery acid!

Scientists have come up with the pH scale for determining the concentration of acid in a solution so we can distinguish between solutions with varying acidity. The pH of a solution can be determined using the following equation:

$$pH = -\log[H^+]$$

where $[H^+]$ is the concentration of H^+ ions, in mol/L. The value of pH itself is unitless, so you can get away with saying that "the pH of this solution is 4.54" without any trouble.

Solutions with a pH less than seven are acidic. Solutions with a pH greater than seven are basic. If a solution has a pH exactly equal to seven, it is neutral.

The Mole Says

Though a solution with a pH of 7.001 is basic from a chemical standpoint, it's very common for people to refer to compounds with a pH near seven as being neutral. Keep in mind that many of the things you would normally consider neutral are either very weak acids or bases.

The following figure shows the acidity of some common substances you may be familiar with.

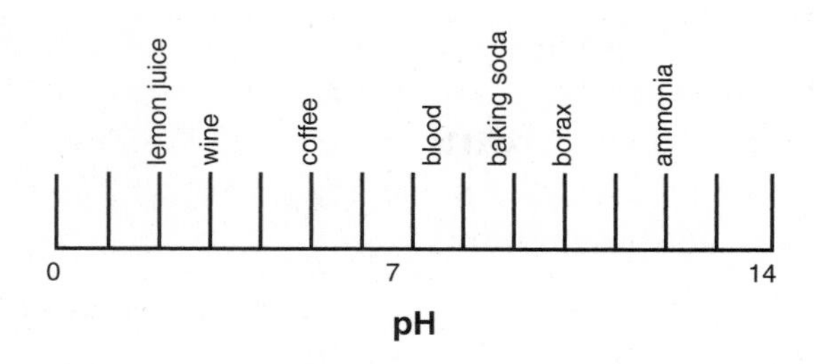

Figure 23.3

The pH scale.

It's possible to discover the pH of a solution by using compounds known as "indicators." Indicators are compounds that change colors as the pH of the solution changes. Probably the two most commonly used indicators are litmus (red in acid and blue in base) and phenolphthalein (pronounced "fee-no-thay-leen," colorless in acid, pink in base).

Finding the pH of a Strong Acid

Strong acids are acids that very nearly completely *dissociate* when you put them into water. That is, almost every molecule of the acid HA that's placed into water breaks up completely to form H^+ and A^- ions. Some common strong acids include HCl, HBr, HI, HNO_3, and $HClO_4$.

> **def•i•ni•tion**
>
> **Strong acids** almost completely break apart in water (a process called **dissociation**) to make H^+ and A^- ions.

Because strong acids completely dissociate in water, the concentration of H^+ in solution is the same as the concentration of the acid you started with. For example, the pH of a 0.00500 M HCl solution would be:

$$-\log(0.00500 \text{ M}) = 2.30$$

> **You've Got Problems**
>
> Problem 2: What is the pH of a solution with a volume of 475 mL containing a total of 1.20 grams of hydrochloric acid?

Finding the pH of a Weak Acid

Weak acids are acids in which most of the molecules don't break apart into H+ and A- ions. As a result, their dissociation is an equilibrium with the following general form.

$$HA \rightleftharpoons H^+ + A^-$$

The equilibrium constant for this expression is given the symbol K_a, which stands for acid-dissociation constant. Some common weak acids include acetic acid and formic acid. You'll only be given K_a values for weak acids, so make sure your brain shifts into "weak acid" gear whenever one is given to you.

def•i•ni•tion

Weak acids are acids that only dissociate to a small degree in water.

As a result, when we place weak acids in water, the H^+ concentration is not the same as the original concentration of the acid. Fortunately, we can use our knowledge of aqueous equilibria (Chapter 22) to find the acid concentration.

Example: What's the pH of a 0.500 M acetic acid solution? K_a ($C_2H_3O_2H$) = 1.75×10^{-5}.

Solution: The equilibrium formed when acetic acid dissociates in water is expressed by the following equation.

$$C_2H_3O_2H \rightleftharpoons C_2H_3O_2^- + H^+$$

To determine the concentration of both, we need to set up the expression for the equilibrium constant (go back to Chapter 22 if this doesn't make sense):

$$K_a = \frac{[C_2H_3O_2^-][H^+]}{[C_2H_3O_2H]}$$

Initially, the concentration of acetic acid is 0.500 M. However, at equilibrium, some of the acetic acid will have dissociated, so the concentration of the acetic acid will have decreased. Since we don't yet know how much acetic acid will have dissociated, we'll express this quantity as "x." As a result, the equilibrium concentration of acetic acid is "0.500 – x" M.

For every molecule of acetic acid that dissociates, one acetate ion and one hydronium ion will be formed. Subsequently, the quantity of each ion formed will be equal to "x." Replacing the quantities in the expression above, we get:

$$1.75x10^{-5} = \frac{[x][x]}{[0.500 - x]}$$

The Mole Says

Another term that's frequently used to describe the strength of an acid is "pKa," defined as: pKa = $-\log(K_a)$

Because acetic acid is a weak acid with a low equilibrium constant, we'll assume that "x" is a small value compared to 0.500 M. We can then simplify the equation above as:

$$1.75x10^{-5} = \frac{[x][x]}{[0.500]}$$

x = 0.00296 M

The concentration of H^+ and $C_2H_3O_2^-$ ions is 0.00296 M. To find the pH, we'll place this value for $[H^+]$ into the equation for pH to get:

$$pH = -\log[H^+]$$

$$pH = -\log(0.00296)$$

$$pH = 2.53$$

You've Got Problems

Problem 3: What's the pH of a 0.750 M formic acid (HCO_2H) solution? K_a (HCO_2H) = 1.77×10^{-4} M.

Finding the pH of Basic Solutions

What's the pH of a 0.0500 M NaOH solution? This is kind of a funny question because when NaOH breaks up in water, it doesn't form H^+ ions. Instead, it forms Na^+ and OH^- ions, which makes it a basic solution.

This poses a problem. After all, the equation for finding pH requires that you have the concentration of H^+ ions, and, as far as we can tell, there are no H^+ ions in a basic solution!

The Mole Says

In pure water, there is a very small quantity of H^+ and OH^- present from the dissociation of water (10^{-7} M, to be exact). However, the quantity of each is so small that you can't detect it simply by drinking it.

Hey, not so fast there, buddy. As it turns out, there are a very small number of H^+ ions in basic solutions. These H^+ ions are formed when water dissociates in the following way.

$$H_2O \rightleftharpoons H^+ + OH^-$$

As a result, whenever water is present (as it must be in an aqueous solution), there will be a few H^+ and OH^- ions—not many, but a few. The equilibrium constant for the dissociation of water has the symbol "K_w". K_w is equal to 10^{-14}.

This turns out to be handy for those of us wanting to find the pH of a basic solution. The reason for this is that the previous equation yields the following equilibrium expression.

$$K_w = [H^+][OH^-]$$

So if we know the concentration of base, we can use this expression to find the concentration of acid.

Example: What is the pH of a 0.0500 M NaOH solution?

Solution: Using the equilibrium expression for water and replacing the values with what we know, we get:

> **You've Got Problems**
>
> Problem 4: What's the pH of a 0.00340 M LiOH solution?

$K_w = [H^+][OH^-]$

$1.00 \times 10^{-14} = [H^+][0.0500]$

$[H^+] = 2.00 \times 10^{-13}$ M

To find pH, we simply use the equation:

$pH = -\log[H^+]$

$pH = -\log[2.00 \times 10^{-13}]$

$pH = 12.7$

This answer indicates a very basic solution, which is what we would expect from a 0.0500 M NaOH solution.

Titrations

I found a bottle in the stockroom of my lab about a year ago. When I opened it, I knew from the burning smell that it was a bottle of nitric acid. However, I had absolutely no idea what the concentration of the acid was. Without knowing the concentration of the acid, it would be difficult to find a use for it.

Fortunately, there's a way to solve this problem. As I remembered from way back in Chapter 18, acid-base reactions occur when an acid and base combine by the following general equation.

$HA + BOH \rightleftharpoons BA + H_2O$

This sparked the following line of reasoning in my brain.

- Every OH^- I add to this solution will neutralize one of the H^+ ions.
- If I keep adding base to my nitric acid, it will eventually turn into neutral water with a pH of 7.00.
- If I keep track of how much base I put into the acid, the quantity of acid when the pH is equal to 7.00 will be equal to the amount of base I've added.

When the solution is perfectly neutral (called the "equivalence point"), the number of moles of acid that I started with will be equal to the number of moles of base that I added to make them neutral. As a result, we get the very handy equation:

$M_aV_a = M_bV_b$

for neutralization reactions at a pH of exactly 7.00. Titration is the process in which neutralization reactions are used to determine the concentration of either an acidic or basic solution.

The Mole Says

In the lab, it's common to determine the point at which a solution has become neutral by using an indicator. When the indicator turns from the color for an acid to the color for a base, the titration has taken place. The use of indicators, however, does cause error because they don't change color at a pH of exactly 7.00. As a result, the point at which an indicator changes color in a titration is the "endpoint" (distinguishing it from the "equivalence point").

Here's how I did my experiment:

I placed 175 mL of nitric acid into a beaker and added 1.00 M NaOH solution to it. I found that the solution was completely neutral after I had added 365 mL of sodium hydroxide to the acid.

Using the earlier equation, M_1 is unknown, V_1 = 175 mL, M_2 = 1.00 M, V_2 = 365 mL, we find that:

$$M_1(175 \text{ mL}) = (1.00 \text{ M})(365 \text{ mL})$$

$$M_1 = 2.09 \text{ M}$$

The nitric acid had a molarity of 2.09 M. Why anybody would make a solution with this molarity, I have no idea, but that's what it was!

You've Got Problems

Problem 5: If it took 245 mL of 0.500 M HCl to neutralize 175 mL of a NaOH solution, what is the concentration of the NaOH solution?

Buffers

Your blood is a very slightly basic solution. Let's consider what happens when you drink a big bottle of soda (pH ~ 3):

- The acid in the soda neutralizes the small quantity of base in your blood.
- The remaining acid in the soda causes the pH of your blood to decrease rapidly to a pH of about 5.
- Because the enzymes in your body don't work well at a pH of 5, you die a horrible death.

def•i•ni•tion

Buffers are solutions that resist changes in pH when acids or bases are added to them. They consist of weak acids and their conjugate bases.

Okay, maybe that's not what happens when you drink a soda. However, if there wasn't something in your blood that stabilizes its pH, it's exactly what would happen.

As it turns out, your blood, like many solutions, is a buffer. *Buffers* are solutions consisting of a weak acid and its conjugate base— these solutions resist changes in pH when either acid or base is added to it.

Let's say that we have a buffered solution that contains acetic acid as its weak acid and sodium acetate as its conjugate base. If we were to add some hydrochloric acid to this solution, the sodium acetate would react with it by the following process:

$$HCl + NaC_2H_3O_2 \rightleftharpoons C_2H_3O_2H + NaCl$$

As you can see, the very strong HCl that's been added to the solution has been converted to acetic acid, which is a weak acid. Because weak acids cause a much smaller disruption in pH than strong acids, the pH of the solution will decrease much less than if it contained no sodium acetate.

Likewise, if we were to add sodium hydroxide to this solution, the acetic acid would react to it by the following process,

$$NaOH + C_2H_3O_2H \rightleftharpoons NaC_2H_3O_2 + H_2O$$

The Mole Says

The quantity of acid or base that a buffer can neutralize before the pH changes greatly is called the "buffering capacity." Buffered solutions with high concentrations of weak acid and conjugate base have higher buffering capacities than those with low concentrations.

Because the strong base NaOH has been converted to the weak base sodium acetate, the pH of the solution won't rise nearly as much as if the acetic acid weren't present in the first place.

We can determine the pH of a buffered solution by using the Henderson-Hasselbalch equation:

$$pH = -\log K_a + \log \frac{[base]}{[acid]}$$

where K_a is the acid dissociation constant of the weak acid in the buffered solution.

Example: What's the pH of a solution that contains 0.100 M acetic acid and 0.200 M sodium acetate? K_a ($C_2H_3O_2H$) = 1.75×10^{-5}.

Solution: Using the Henderson-Hasselbalch equation, we find that:

$$pH = -\log K_a + \log \frac{[base]}{[acid]}$$

$$pH = -\log(1.75x10^{-5}) + \log \frac{[0.200]}{[0.100]}$$

$$pH = 4.76 + 0.301$$

$$pH = 5.06$$

You've Got Problems

Problem 6: Determine the pH of a solution containing 0.50 M formic acid and 0.75 M lithium formate. The K_a value of formic acid is 1.8×10^{-4}.

The Least You Need to Know

- The three types of acids and bases are Arrhenius acids and bases, Brønsted-Lowry acids and bases, and Lewis acids and bases.
- The pH scale is commonly used to describe the acidity of solutions.
- A titration is the use of a neutralization reaction to find the concentration of an acid or a base.
- Buffers are mixtures of weak acids and their conjugate bases that resist changes in pH.

Chapter 24

Electrochemistry

In This Chapter

- Oxidation states
- Redox reactions
- Voltaic cells
- The Nernst equation
- Electrolytic cells

When I was in first grade, a "friend" of mine told me that something neat would happen if I touched the two terminals of a 9-volt battery to my tongue. When I asked him what would happen, he told me that it was a secret, but that it would be a lot of fun. I very quickly learned about electrochemical reactions in a way that was somewhat less than fun, what with the pain in my tongue and all.

I guess that's a roundabout way of saying that electrochemistry is all around us. From the batteries we shock ourselves with as small children to the electroplating reactions that give cheap jewelry a very thin veneer of gold, electrochemistry is a part of life. It's time we learned more about it!

Oxidation States

Before we can talk about how electrochemical reactions occur, we need to work through the basics. In electrochemistry, nothing is more basic than the concept of oxidation states.

Oxidation states describe the charges that the atoms in chemical compounds are considered to have. In simple ionic compounds, the oxidation states of cations and anions are the same as their charges (for example, in NaCl, the oxidation state of sodium is +1 and the oxidation state of chlorine is –1). However, for covalent compounds and complex ions, we need to make use of some rules to help us out.

For materials other than ionic compounds, things are a little more complicated. To find these oxidation states, follow these rules.

1. The oxidation states of pure elements are zero. For example, "Fe" has a zero oxidation state, as do each of the chlorine atoms in "Cl_2."
2. The oxidation state of the most electronegative element in a compound is the same as it would normally have if it were an anion. For example, in BF_3, we assume that fluorine has an oxidation state of –1 because it's more electronegative than boron.
3. The oxidation state of hydrogen is normally +1. The exception comes if it is bonded to a metal, in which case the oxidation state is –1. For example, in CH_4, hydrogen has an oxidation state of +1, while in NaH, it has an oxidation state of –1.
4. In any neutral chemical compound, the sum of the oxidation states of all elements is zero. We can use this rule to figure out what the oxidation state of oxygen is in H_2O_2—because each hydrogen atom has a charge of +1, each oxygen atom must have a charge of –1 for the sum of the oxidation states to equal zero.
5. In all polyatomic ions, the sum of the oxidation states of all elements is equal to the charge of the ion. For example, in NH_4^+, the oxidation state of each hydrogen atom is +1, for a total of +4. The overall charge of the ion is +1, making the charge of nitrogen +1 – 4 = –3.

def•i•ni•tion

The **oxidation state** (a.k.a. "oxidation number") of an atom is the charge that the atom is considered to possess in a chemical compound.

You've Got Problems

Problem 1: What are the oxidation states of the elements in the following compounds?
(a) PBr_3 (b) NaOH (c) H_2SO_4

Oxidation and Reduction

During electrochemical processes, atoms gain or lose electrons. As a result, their oxidation numbers will change during the course of an electrochemical reaction. When an atom loses electrons, causing it to have a more positive oxidation state, it is said to have been *oxidized*. When an atom gains electrons, causing it to have a more negative oxidation state, it has been reduced. Reactions in which oxidation and reduction occur are referred to as "oxidation-reduction" reactions, or more commonly *redox reactions*. To keep oxidation and reduction clear in your mind, use the following lion-related phrase: LEO goes GER! LEO stands for Lose Electrons = Oxidation and GER stands for Gain Electrons = Reduction.

def•i•ni•tion

An atom has been **oxidized** if it loses electrons and reduced if it gains electrons. Reactions in which the oxidation states of the elements change are called **redox reactions**.

Let's take a look at a sample redox reaction and determine which elements have been oxidized and reduced.

$$2\ Na + ZnCl_2 \rightleftharpoons Zn + 2\ NaCl$$

- On the reactants side of the equation, sodium has an oxidation state of zero because it is a pure element. On the products side of the equation, it has an oxidation state of +1. We say that sodium has been oxidized because its charge has been made more positive (0 to +1) by the loss of electrons.
- On the reactants side of the equation, zinc has a +2 oxidation state because it is bonded to two chloride ions. On the products side of the equation, it has an oxidation state of zero because it is a pure element. We say that zinc has been reduced because the charge has been made less positive (+2 to 0) by the gain of electrons.
- Chlorine has a –1 oxidation state on both sides of the equation. It has been neither oxidized nor reduced.

In redox reactions, you can't have one element oxidized without another element having been reduced. After all, it's the movement of electrons that causes both oxidation and reduction, so the total number of electrons removed from one thing via oxidation is the same as the number of electrons added to another via reduction. Because elements that gain electrons have pulled them away from the elements that have been oxidized, they are considered "oxidizing agents" or "oxidants." Elements that have lost electrons have given them to elements that are reduced, so they are called "reducing agents" or "reductants." For our earlier example, sodium is the reducing agent (it caused zinc to be reduced), and zinc chloride is the oxidizing agent (it caused sodium to be oxidized).

You've Got Problems

Problem 2: Determine the oxidizing and reducing agents in the following electrochemical process:

$CH_4 + 2\ O_2 \rightleftharpoons CO_2 + 2\ H_2O$

Balancing Redox Reactions

In this section, we're going to discuss balancing redox reactions. Now, you may be asking yourself, "Why do I need to learn how to balance redox reactions? After all, I learned how to balance equations way back in Chapter 18! What's the deal?"

First of all, how dare you take that tone with me! (Just kidding.) Second, redox reactions are sometimes harder to balance than other reactions.

Redox reactions are balanced using the half-reaction method. Though there are a lot of steps in the half-reaction method, you'll soon get the hang of it!

Example: Balance the following redox reaction.

$$Al + MnO_2 \rightleftharpoons Mn + Al_2O_3$$

Solution: Before we solve the problem, you may have noticed that this redox reaction isn't really all that difficult to balance using the method we learned in Chapter 18. I'm giving you this reaction because I want to start off with something simple to get you used to the process.

To balance redox reactions with the half-reaction method, follow these steps.

Step 1

Break the unbalanced equation into two smaller equations called half-reactions. The first half-reaction will follow the oxidation of the element that lost electrons, and the second will follow the reduction of the element that gained electrons.

Because aluminum is oxidized in this reaction, the oxidation half-reaction is:

$$Al \rightarrow Al_2O_3$$

The half-reaction for the reduction of manganese is:

$$MnO_2 \rightarrow Mn$$

Step 2

In each half-reaction, balance the elements that are oxidized or reduced. After you're done with that, balance any elements other than oxygen or hydrogen.

For the oxidation half-reaction, we balance the aluminum.

$$2\ Al \rightarrow Al_2O_3$$

For the reduction half-reaction, we don't need to balance anything, as there is only one manganese atom on both the reactant and product side.

Step 3

Balance the oxygen atoms by adding H_2O.

For the oxidation half-reaction, we do this by adding three water molecules to the reactant side of the equation.

$$2\ Al + 3\ H_2O \rightarrow Al_2O_3$$

For the reduction half-reaction, we add two water molecules to the product side of the reaction.

$$MnO_2 \rightarrow Mn + 2\ H_2O$$

The Mole Says

We add water to these equations because many redox reactions take place in water. Even for reactions that don't take place in water, the water molecules we add in this step will eventually cancel each other out.

Step 4

Balance the hydrogen atoms for both half-reactions by adding H^+ ions.

This is done for each half-reaction in the following way.

$$2\ Al + 3\ H_2O \rightarrow Al_2O_3 + 6\ H^+$$

$$MnO_2 + 4\ H^+ \rightarrow Mn + 2\ H_2O$$

Step 5

Add electrons so the total amount of charge on both sides of the reaction is neutral.

Because the only charges present in each of these half-reactions are caused by H^+ (this, incidentally, isn't always the case), we'll add electrons to neutralize these ions.

$$2\ Al + 3\ H_2O \rightarrow Al_2O_3 + 6\ H^+ + 6\ e^-$$

$$MnO_2 + 4\ H^+ + 4\ e^- \rightarrow Mn + 2\ H_2O$$

Step 6

Multiply the coefficients for both half reactions so the number of electrons in each is the same.

In our example, there are six electrons in the oxidation half-reaction and four in the reduction half-reaction. To make both numbers of electrons the same, we can multiply the coefficients in the first half-reaction by two (to get a total of twelve electrons in the first reaction) and the second by three (to get a total of twelve electrons in the second reaction).

$$4\ Al + 6\ H_2O \rightarrow 2\ Al_2O_3 + 12\ H^+ + 12\ e^-$$

$$3\ MnO_2 + 12\ H^+ + 12\ e^- \rightarrow 3\ Mn + 6\ H_2O$$

Step 7

Add the two half-reactions together.

$$4\ Al + 6\ H_2O + 3\ MnO_2 + 12\ H^+ + 12\ e^- \rightarrow 2\ Al_2O_3 + 12\ H^+ +$$
$$12\ e^- + 3\ Mn + 6\ H_2O$$

Step 8

Cancel out any terms that are present in equal amounts on both sides of the resulting equation.

This leaves:

$$4\ Al + 3\ MnO_2 \rightleftharpoons 2\ Al_2O_3 + 3\ Mn$$

… which is the correct answer!

You've Got Problems

Problem 3: Balance the following redox reaction:

$As_2O_3 + NO_{3^-} \rightleftharpoons H_3AsO_4 + NO$

Voltaic Cells

So, who uses electrochemistry, anyway? As it turns out, you do. One of the main uses of electrochemistry is in the batteries you use to power your portable radio, pacemaker, wristwatch, and cordless electric razor (assuming you have these items, that is). Someday soon, your car will even run on battery power! Because batteries are so important, let's learn more about them.

Introduction to Voltaic Cells

As it turns out, another fancy word for "battery" is "voltaic cell" (sometimes known as a "galvanic cell"). An example of a voltaic cell is shown in Figure 24.1:

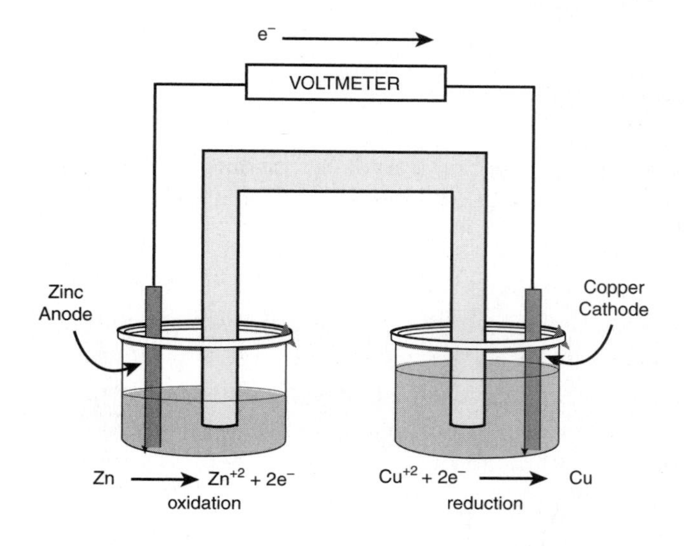

Figure 24.1

A voltaic cell in which zinc is oxidized and copper is reduced.

Before we can really get down to the nuts and bolts of how voltaic cells work, we need to understand some background information about them:

- The pieces of metal that are dipped into the solution in this diagram are called electrodes. Electrodes are connected to one another by a wire on one side and through the solution on the other side.

- The electrode at which oxidation occurs is called the anode, and the electrode at which reduction occurs is called the cathode. In this diagram, zinc is the anode and copper is the cathode.
- The U-shaped tube between the two beakers in this diagram is called a "salt bridge." Salt bridges contain an electrolyte solution (a solution that conducts electricity) that won't react with any of the other chemicals in the cell. Salt bridges are required because the charge carriers in a voltaic cell are ions, and for electricity to be conducted, the two sides of the cell need to be connected.
- The term "half-cell" refers to the process that takes place at each of the electrodes. Every battery has two half-cells because all batteries have two electrodes.
- The electrochemical reactions that occur in a voltaic cell are written using the following shorthand: The oxidation reaction is written first—in our cell, zinc is converted to Zn^{+2}, so the reaction is written as $Zn|Zn^{+2}$ (the single line between the Zn and Zn^{2+} indicates what happens to the zinc before and after the electrochemical process occurs). The reduction reaction is written second—in our cell, Cu^{2+} is converted to copper, so the reaction is written as $Cu^{2+}|Cu$. To write the overall process for the cell, write these reactions in order, with a double vertical line separating them: $Zn|Zn^{+2}||Cu^{+2}|Cu$.

Standard Electrode Potentials

As we mentioned earlier, the batteries used in everyday electronic devices include voltaic cells. However, you can't just put any two solutions into a voltaic cell to make a useful battery.

One of the most important factors when considering how batteries work is their voltage. Voltage is just a measure of how forcefully electrons are moved from one place to another and is defined as the amount of energy given off by a spontaneous electrochemical process or the amount of energy needed to cause a nonspontaneous redox reaction to take place. Wouldn't it be useful to figure out how to calculate the voltage of a battery?

The Mole Says

The little ° above the cell potentials in this equation indicates that the cell is running under standard conditions. Namely, the concentrations of the reactants and products are exactly 1 M for solutions and 1 atm for gases.

You bet it would! In order to calculate the total standard cell potential, you use the following equation.

$$E^\circ_{cell} = E^\circ_{oxidation} + E^\circ_{reduction}$$

This means that to find the potential of a cell, you need to add the potentials of the reactions that take place at the anode and cathode half-cells. This, of course, requires that we know what the half-cell potentials are.

Fortunately, nice chemists have already made big tables of half-cell potentials that we can use. Let's take a look at some standard reduction potentials.

Standard Reduction Potential (V)	Reduction Half-Reaction
2.87	$F_{2(g)} + 2\ e^- \rightarrow 2\ F^-_{(aq)}$
1.99	$Ag^{+2}_{(aq)} + e^- \rightarrow Ag^+_{(aq)}$
1.82	$Co^{+3}_{(aq)} + e^- \rightarrow Co^{+2}_{(aq)}$
0.80	$Ag^+_{(aq)} + e^- \rightarrow Ag_{(s)}$
0.77	$Fe^{+3}_{(aq)} + e^- \rightarrow Fe^{+2}_{(aq)}$
0.52	$Cu^+_{(aq)} + e^- \rightarrow Cu_{(s)}$
0.34	$Cu^{+2}_{(aq)} + 2\ e^- \rightarrow Cu_{(s)}$
0.002	$H^+_{(aq)} + 2\ e^- \rightarrow H_{2(g)}$
–0.28	$Ni^{+2}_{(aq)} + 2\ e^- \rightarrow Ni_{(s)}$
–0.44	$Fe^{+2}_{(aq)} + 2\ e^- \rightarrow Fe_{(s)}$
–0.76	$Zn^{+2}_{(aq)} + 2\ e^- \rightarrow Zn_{(s)}$
–1.66	$Al^{+3}_{(aq)} + 3\ e^- \rightarrow Al_{(s)}$
–2.71	$Na^+_{(aq)} + e^- \rightarrow Na_{(s)}$

You may have noticed that this chart lists only reduction potentials, not oxidation potentials. Fortunately, you can find the oxidation potential of a half-reaction by reversing the sign of the reduction potential. For example, by reversing the first entry in the previous chart, we find that the standard oxidation potential for F^- is –2.87 V.

The Mole Says

Electrochemical processes are only spontaneous if the cell potential is positive.

Now that we know how to find the potential of a voltaic cell, let's do it for the cell we discussed earlier: $Zn \mid Zn^{+2} \mid \mid Cu^{+2} \mid Cu$.

The following two half-reactions take place.

$Zn \rightarrow Zn^{+2} + 2\ e^-$ oxidation

$Cu^{+2} + 2\ e^- \rightarrow Cu$ reduction

You've Got Problems

Problem 4: Determine the standard cell potential for the following voltaic cell:

$Al \mid Al^{+3} \mid \mid Fe^{+2} \mid Fe$

To find the overall cell potential, we simply need to add the half-cell potentials. For the oxidation of zinc to Zn^{+2}, the half-cell potential is the same as for the reduction of zinc, except with the sign changed from negative 0.76 V to positive 0.76 V. For the reduction of Cu^{+2} to pure copper, it's 0.34 V. When you add them up using the equation for standard cell potential, you find that:

$$E^o_{cell} = E^o_{oxidation} + E^o_{reduction}$$

$$= 0.76\text{ V} + 0.34\text{ V}$$

$$= 1.10\text{ V}$$

The Nernst Equation

The preceding calculations are really handy when all of the reactions take place under standard conditions. However, sometimes these reactions don't occur under standard conditions. When this happens, it's time to call in the Nernst equation:

$$E = E^o - \frac{0.0591}{n}\log Q$$

In this equation, E is the cell potential under the conditions given, E° is the standard cell potential, n is the number of electrons transferred in the reaction, and Q is the reaction quotient.

The Mole Says

The reaction quotient for the generic process aA + bB ⇌ cC + dD is:

$$Q = \frac{[C_o]^c[D_o]^d}{[A_o]^a[B_o]^b}$$

Where C_o is the initial molarity of C, D_o is the initial molarity of D, and so on. For gaseous equilibria, partial pressures should be used in lieu of molarities.

Example: For the cell $Zn|Zn^{+2}||Cu^{+2}|Cu$, what is the cell potential if the concentration of Zn^{+2} is 2.5 M and the concentration of Cu^{+2} is 0.75 M?

Solution: In the previous section, we found that E^o_{cell} for this process was 1.10 V. However, before using the Nernst equation, it is necessary to figure out what values we should use for all the variables.

For this process, n = 2 because two electrons are transferred from Zn to Cu^{+2} whenever this reaction takes place.

To find the reaction quotient, we need to write out the equation for the entire process.

$$Zn_{(s)} + Cu^{+2}_{(aq)} \rightleftharpoons Zn^{+2}_{(aq)} + Cu_{(s)}$$

$$Q = \frac{[Zn^{+2}][Cu]}{[Zn][Cu^{+2}]}$$

We don't need to include solids in our equilibrium expressions because they don't have any concentration. As a result, Q is:

$$Q = \frac{[Zn^{+2}]}{[Cu^{+2}]} = \frac{2.5M}{0.75M} = 3.3$$

Putting all of these values into the Nernst equation, we find that:

$$E = 1.10V - \frac{0.0591}{2}\log 3.3$$

E = 1.10 V – 0.015 V

E = 1.08 V

Electrolytic Cells

If you've ever watched the home shopping channels, you know that electroplating is big business. For those of you who don't know what I'm talking about, electroplating is a process in which a very thin coating of a precious metal is placed over a very cheap metal. Common examples of electroplated materials include cheap jewelry, dinnerware, and just about anything bought from TV shopping networks.

Electroplating is made possible by electrolytic cells. Electrolysis is a process by which a current is forced through a cell in order to make a nonspontaneous electrochemical change occur. For example, by forcing electricity through a cell, we can force electrochemical reactions with negative cell potentials to occur. An electrolytic cell is shown in Figure 24.2.

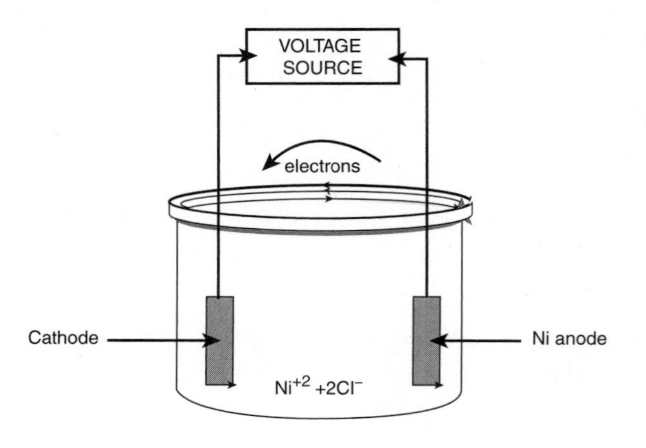

Figure 24.2

An electrochemical cell for plating nickel on stainless steel.

In the electrochemical cell in Figure 24.2, the following process takes place.

1. $NiCl_2$ dissolves to form Ni^{+2} and 2 Cl^- ions.
2. When the cell is turned on, Ni^{+2} ions move toward the cathode, and Cl^- ions move toward the anode.
3. At the cathode, the Ni^{+2} ions are reduced to form a thin layer of nickel. Typically, the item to be plated is used as the cathode.
4. Meanwhile, the nickel anode will oxidize, generating more Ni^{+2} ions to replace those that were plated to the silverware.

During this process, the anode will eventually disappear, having been electroplated onto the cathode.

The Least You Need to Know

- The oxidation state of an element is equal to the amount of charge it can be considered to have in a compound.
- Oxidation occurs when an element loses electrons, and reduction occurs when an element gains electrons.
- Redox reactions occur whenever a chemical change is accompanied by a change in oxidation states of the elements present.
- In voltaic cells, electrons are transferred from one half-cell to another, resulting in the flow of electricity.
- The Nernst equation is used for determining the voltaic cell potentials under nonstandard conditions.
- In electrolytic cells, electricity is forced through a solution to make a non-spontaneous electrochemical reaction occur. This process is commonly used in electroplating.

Chapter 25

Organic Chemistry

In This Chapter

- Hydrocarbons
- Isomerism
- Functional groups
- Basic organic reactions

Organic chemistry is a lot of fun. When you understand how organic molecules react with one another, you truly get a feel for how rich and exciting the field of chemistry really is. The first time I took organic chemistry, I was hooked!

Unfortunately, organic chemistry is much too big a subject to discuss with any detail in a general chemistry course. As a result, students usually don't understand why they have to learn it or how it fits together with the rest of chemistry. Because I realize that this is a problem with how organic chemistry is taught in a first-year course, I've focused my attention on topics you're likely to see on a test. It's my hope you'll decide to take a course devoted solely to organic chemistry so you can see how cool it really is.

So sit back and relax as we take a whirlwind tour through the world of organic chemistry.

What Is Organic Chemistry?

After the big buildup I just gave organic chemistry, you may be wondering what it is. Organic chemistry is the study of carbon-containing molecules. Most compounds that contain carbon are referred to as organic molecules; the only common carbon-containing molecules that aren't commonly referred to as "organic" are the oxides of carbon (CO, CO_2) and carbonates.

Up to this point, we've been talking about reactions involving many of the elements in the periodic table, so you might think that organic chemistry isn't that important—after all, carbon is only one element. However, organic molecules usually contain hydrogen, and many contain oxygen, nitrogen, the halogens, sulfur, phosphorus, and a variety of other elements. Carbon also likes to form long chains and rings—as a result, millions of *organic compounds* are known, and there is an infinite number of organic compounds that can be formed. Not too shabby for just one element!

def•i•ni•tion

Organic compounds consist of carbon-containing compounds, with the exception of oxides (CO, CO_2) and carbonates (molecules containing the CO_3^{2-} ion).

Hydrocarbons

Hydrocarbons are molecules that contain only carbon and hydrogen. Even with only these two elements, there are still a huge variety of compounds that can be formed. Because there are so many possible molecules, the naming system that has evolved to tell them apart from each other is fairly complex. As a result, it's frequently not only fun to do organic chemistry, but also fun to say the names of each compound very quickly to confuse people.

Alkanes

Alkanes are hydrocarbons in which all bonds to carbon are single bonds. Because carbon bonds to four different atoms in alkanes, these molecules are referred to as "saturated hydrocarbons."

The naming system used for organic molecules is based on the names of alkanes in which all of the carbon molecules are arranged in straight chains. The first eight alkanes are shown in Figure 25.1.

def•i•ni•tion

Alkanes (also called "saturated hydrocarbons") are hydrocarbons that contain only single carbon-carbon bonds and carbon-hydrogen bonds.

Number of carbon atoms	Name	Formula	Structure
1	methane	CH_4	H H - C - H H
2	ethane	C_2H_6	H H H - C - C - H H H
3	propane	C_3H_8	
4	butane	C_4H_{10}	
5	pentane	C_5H_{12}	
6	hexane	C_6H_{14}	
7	heptane	C_7H_{16}	
8	octane	C_8H_{18}	

Figure 25.1

The first eight straight-chain hydrocarbons. Make sure you memorize the names of these compounds because you'll need them later!

In Figure 25.1, all of the atoms for methane and ethane are shown, but for propane through octane, only straight lines are drawn. This is a common shorthand method of showing the structure of organic molecules. It's assumed that the intersections of all points correspond to carbon atoms, as do the ends of lines. Hydrogen atoms are added to this structure so that all carbon atoms have a total of four bonds. To completely draw out pentane, for example, follow the steps in Figure 25.2.

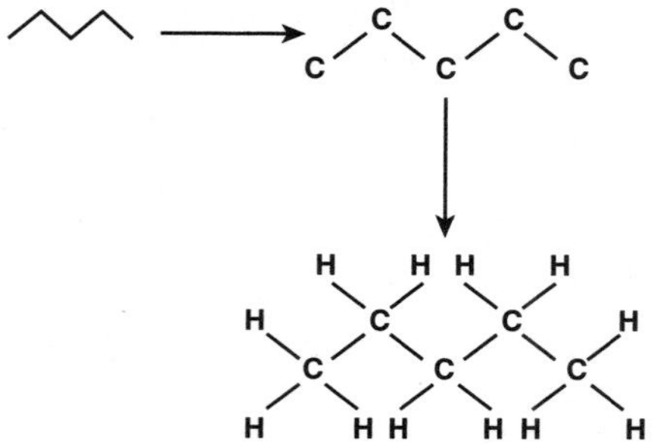

Figure 25.2

A pentane molecule.

To name alkanes, follow these rules.

- The name of a chemical compound is based on the longest unbroken chain of carbon atoms in a molecule. For example, the molecule in Figure 25.3 is said to be a "*hex*ane" because the longest carbon chain has six atoms.

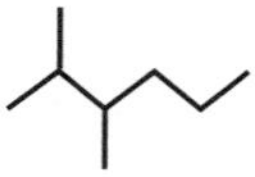

Figure 25.3

The name of this molecule will contain "hexane" because the longest carbon chain has six atoms.

- Any group that hangs off of the longest chain is named based on how many carbon atoms it contains. In the preceding diagram, each of the two groups hanging off of the longest chain has one carbon atom, so they're referred to as "methyl" groups. If a group has two carbon atoms, it's called an "ethyl" group, and so on. Substituents such as these are called "alkyl groups" to indicate that they have the same basic name as their parent alkanes, but with a "-yl" ending.
- Because there are many positions where a group can be located on the chain, we have to indicate which carbon atom in the chain it's attached to. To do this, we number the carbon atoms from each end of the chain such that the alkyl group positions have the smallest possible numbers. In our earlier example, we can number the chain in two possible ways, as shown here.

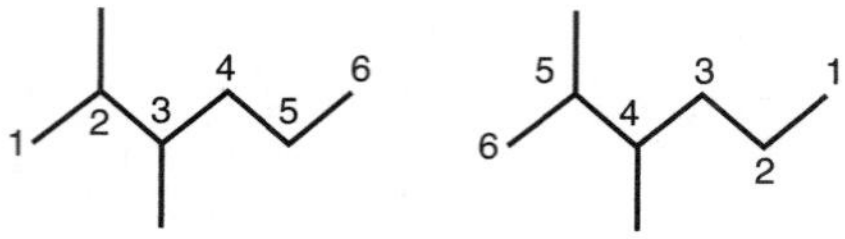

Figure 25.4

The numbering scheme on the left is correct because the methyl (CH_3) groups are on the second and third carbon atoms in the chain, rather than the fourth and fifth carbon atoms if you number it in the opposite direction.

- If there is more than one of a substituent, use the prefix di- for "two," tri- for "three," tetra- for "four," and so on to indicate how many there are. Before the prefixes, indicate the carbon that each is stuck to. In our example, the molecule is referred to as "2,3-dimethylhexane" because one methyl group is on the second carbon and the other is on the third. Take care to include a hyphen between the numbers and the name.
- If there is more than one of a particular substituent (e.g. "methyl"), write them in alphabetical order, regardless of their position on the chain or prefixes. For example, the molecule in Figure 25.5 is called "3-ethyl-2,4-dimethyloctane":

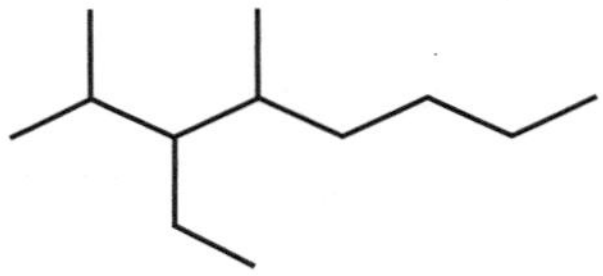

Figure 25.5

A 3-ethyl-2,4-dimethyloctane molecule.

You've Got Problems

Problem 1: Draw the structure of 4-ethyl-2-methylhexane.

Alkenes and Alkynes

Alkenes and alkynes are both unsaturated hydrocarbons, which means that there's at least one carbon-carbon multiple bond. Alkenes have at least one carbon-carbon double bond, and alkynes have at least one carbon-carbon triple bond.

Alkenes are named in much the same ways as alkanes, except that the ending of the molecule is "-ene" instead of "-ane." Additionally, a number is added before the name of the longest chain to indicate the position of the double bond.

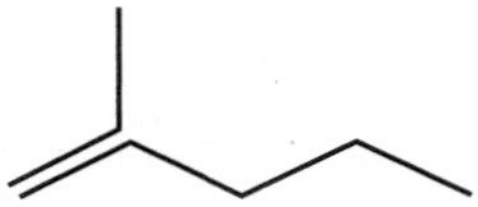

Chemistrivia

You may have heard of "unsaturated" and "polyunsaturated" vegetable oils. Unsaturated vegetable oils contain one carbon-carbon double bond, while polyunsaturated oils contain more than one C=C bond.

Figure 25.6

A 2-methyl-1-pentene molecule.

The molecule in Figure 25.6 is referred to as "2-methyl-1-pentene" to indicate that the first atom is where the double bond begins and the second atom contains a methyl group.

Likewise, alkynes are named such that the ending of the molecule is "-yne." The following figure shows 4-methyl-2-hexyne.

Figure 25.7

A 4-methyl-2-hexyne molecule.

You've Got Problems

Problem 2: Draw 4-ethyl-2-methyl-2-hexene.

Cyclic Hydrocarbons

Carbon atoms frequently form rings. If the rings contain only carbon-carbon single bonds, they're referred to as cycloalk*a*nes. If they contain at least one carbon-carbon double bond, they're called cycloalk*e*nes. If they contain a carbon-carbon triple bond, they're called cycloalk*y*nes. These molecules are named in the same way that straight-chain alkanes, alkenes, and alkynes are named, except that the carbon atoms in the ring are numbered such that the substituents have the smallest possible numbers. An example of a cyclic cycloalkane would be 1,2-diethylcyclopentane.

Figure 25.8

A 1,2-diethylcyclopentane molecule.

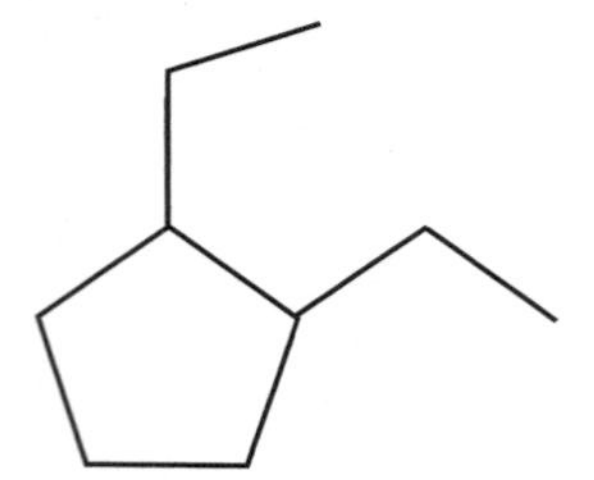

You've Got Problems

Problem 3: Draw 1-ethyl-3-methylcyclohexane.

Though cyclic molecules with three atoms can be formed, these molecules typically aren't very stable. Recall from Chapter 10 that sp^3-bonded atoms are most stable when the bond angles are 109.5°. However, in cyclopropane, the bond angle is forced to a very small 60°, which puts a lot of strain on the ring. This ring strain is called, straightforwardly enough, "ring strain." Generally, five- and six-membered rings have the least ring strain and are most commonly formed.

Aromatic Hydrocarbons

Aromatic hydrocarbons are cyclic molecules that are drawn with alternating carbon-carbon single and double bonds. Probably the best-known aromatic hydrocarbon is benzene.

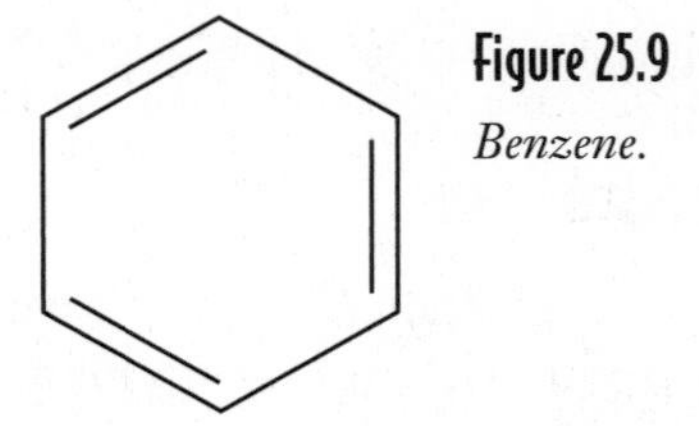

Figure 25.9
Benzene.

The electrons in benzene's double bonds are said to be "delocalized" because they can travel around the entire ring. This can be seen more clearly by examining benzene's resonance structures.

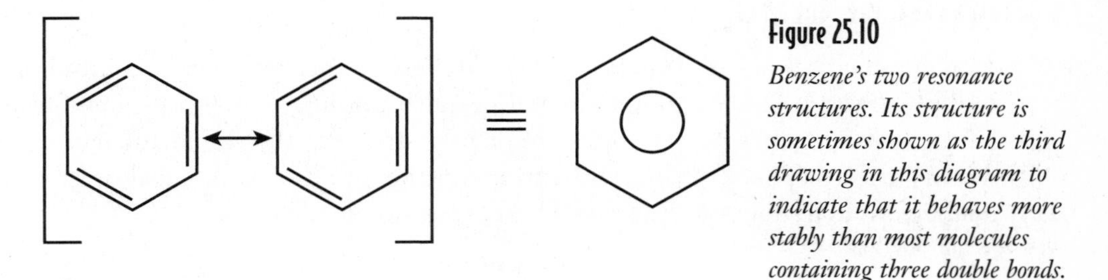

Figure 25.10
Benzene's two resonance structures. Its structure is sometimes shown as the third drawing in this diagram to indicate that it behaves more stably than most molecules containing three double bonds.

Because the electrons in benzene's three double bonds are delocalized, benzene is an unusually stable molecule.

Isomers

Just because two molecules have the same molecular formula doesn't mean they have the same structural formulas. Different molecules that have the same molecular formula are known as *isomers*.

def•i•ni•tion

Isomers are different molecules that have the same formulas.

There are two main types of isomerism, as we'll see next.

Constitutional Isomerism

Constitutional isomers are molecules that have the same formulas, but differ in the order that the atoms are connected to each other. An example of constitutional isomerism is shown here.

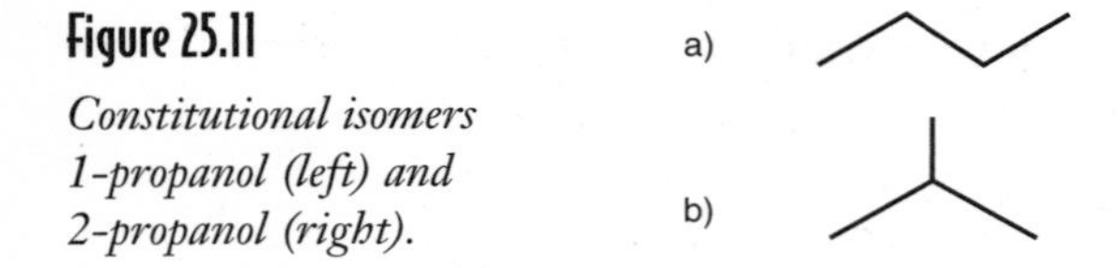

Figure 25.11
Constitutional isomers 1-propanol (left) and 2-propanol (right).

Stereoisomerism

Stereoisomers are molecules that have the atoms bonded in the same order, but with different spatial orientations. To imagine what a stereoisomer looks like, hold your hands in front of you. Though each has four fingers and a thumb connected to your palm, your hands are different from one another in that they can't fit into the same glove.

Functional Groups

As mentioned earlier in the chapter, there are many other elements present in organic molecules besides carbon and hydrogen. A quick warning: Many organic molecules have common names that are more frequently used than the systematic names we're discussing in this book. When the common names are used with equal or greater frequency, I'll write them in parentheses after the systematic name.

The Mole Says

In this diagram, you'll see the letter "R" shows up a lot when I'm introducing new functional groups. "R" is a way of representing a generic organic group. For example, if I were to write R-Br, this could mean CH_3Br, C_6H_5Br, or anything else organic with a –Br stuck to it.

The following figure shows the most common organic functional groups.

Functional group	General Formula	Example	Name
alkyl halide	R-X	CH_3Br	bromomethane (methyl bromide)
alcohol	R-OH	CH_3OH	methanol
ether	R-O-R' (R groups not necessarily the same)	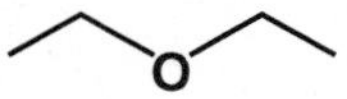 O	diethyl ether
aldehyde	O ‖ R—C—H	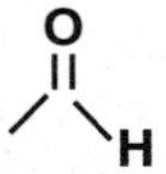 O ‖ H	ethanal (acetaldehyde)
ketone	O ‖ R / C \ R'	 O ‖	propanone (acetone, dimethyl ketone)
carboxylic acid	O ‖ R—C—OH	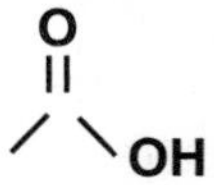 O ‖ OH	ethanoic acid (acetic acid)
ester	O ‖ R / C \ O / R'	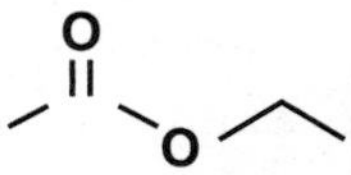 O ‖ O	ethyl ethanoate (ethyl acetate)
amine	R—N—R'' with R' on N	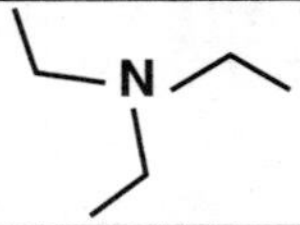 N	triethylamine

Figure 25.12

The most common organic functional groups and their names.

Some facts about these organic groups that you may find interesting or useful.

- What we commonly think of as "alcohol" or "grain alcohol" is known chemically as ethanol. "Rubbing alcohol" is 2-propanol (isopropyl alcohol), and "wood alcohol" is methanol.
- Alcohols are known as "primary alcohols" if the –OH group is on the first carbon. Likewise, they are "secondary alcohols" if the –OH is located on carbons in the middle of the chain, and "tertiary alcohols" if the OH is located on a carbon that's bonded to three other carbon atoms. For example, 1-propanol is a primary alcohol, 2-propanol is a secondary alcohol, and 2-methyl-2-propanol (also called t-butyl alcohol) is a tertiary alcohol.

- The hydrogen in the "–OH" group in carboxylic acids is weakly acidic—one common example is ethanoic acid (acetic acid), which is found in vinegar.
- Esters frequently have very pleasant fruity or floral smells, and are commonly found in perfumes.
- Amines usually smell awful. For example, triethylamine smells strongly of dead fish.

Organic Reactions

Now that we're familiar with organic chemicals, it's time to learn about their reactions. The most comprehensive organic chemistry manual I own lists only the fundamentals of organic chemistry, and it's 1,495 pages long. Because nobody wants to read a 1,500-page chapter, the following is a very abbreviated list of organic reactions.

Addition Reactions

Alkenes and alkynes frequently react such that the multiple bonds are replaced by single bonds with other elements. One such type of reaction is called a "hydrogenation reaction" because hydrogen is added to an alkene.

Figure 25.13

The hydrogenation of ethene (ethylene) to form ethane.

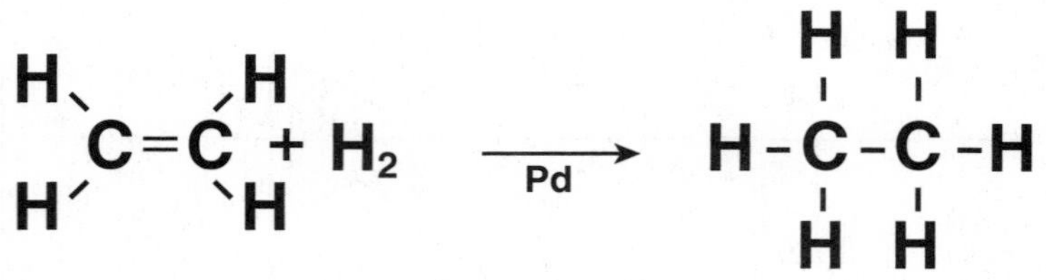

$$\mathrm{H_2C{=}CH_2 + H_2 \xrightarrow{Pd} H{-}CH_2{-}CH_2{-}H}$$

Halogenation reactions occur when alkenes or alkynes react with halogen molecules.

Figure 25.14

The bromination of ethene to form 1,2-dibromoethane.

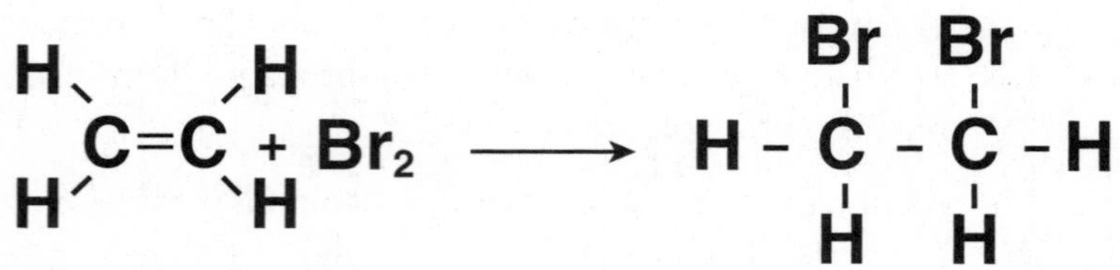

$$\mathrm{H_2C{=}CH_2 + Br_2 \longrightarrow H{-}CHBr{-}CHBr{-}H}$$

When hydrogen halides react with alkenes or alkynes, the halogen atom bonds with the carbon atom with fewer hydrogen atoms. An example of this is the reaction of HBr with propene to form 2-bromopropane.

Figure 25.15

A 2-bromopropane molecule.

Likewise, when water reacts with alkenes or alkynes, alcohols are formed. As with the preceding reaction, the –OH group always bonds with the carbon atom, which has fewer hydrogen atoms.

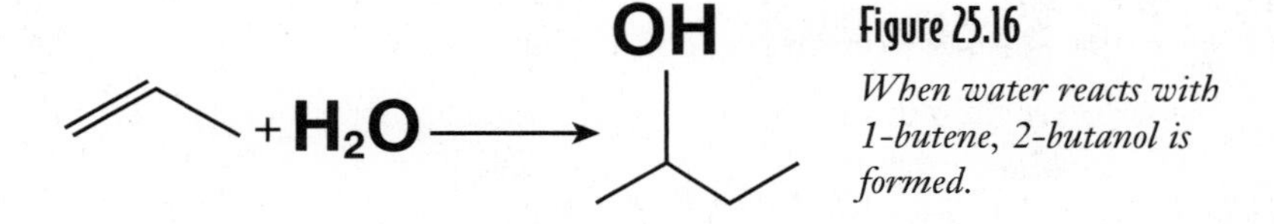

Figure 25.16

When water reacts with 1-butene, 2-butanol is formed.

Free-Radical Substitution Reactions

Alkyl halides can be formed when alkanes react with halogens in the presence of light. This process takes place via a free radical process. Free radicals are highly reactive atoms or groups of atoms with an unpaired electron.

Free-radical reactions typically take place in a three-step chain reaction. We'll see how this works for the reaction of chlorine with ethane to form chloroethane.

light

$Cl_2 + CH_3CH_3 \rightleftharpoons CH_2ClCH_3 + HCl$

Step 1

Initiation: In the initiation step, the reactive species is generated when a halogen is broken apart with light.

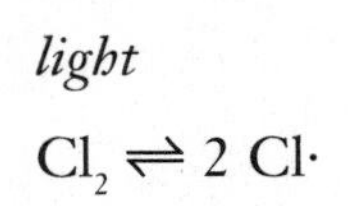

light

$Cl_2 \rightleftharpoons 2\ Cl\cdot$

The Mole Says

That dot next to some of these chemicals isn't a typo! It represents the unpaired electron in a free radical and responsible for their very high reactivity.

Step 2

Propagation: The propagation step involves the chlorine radicals reacting with the alkane molecule to form alkane radicals. In turn, the alkane radicals can react with halogen molecules to form more alkane radicals. The propagation steps for this reaction are shown here.

$$Cl\cdot + CH_3CH_3 \rightleftharpoons HCl + \cdot CH_2CH_3$$

$$\cdot CH_2CH_3 + Cl_2 \rightleftharpoons CH_2ClCH_3 + Cl\cdot$$

Step 3

The last steps in a free radical reaction are the termination steps. The termination steps occur anytime that two free radicals combine with one another, as this removes reactive species from the reaction mixture. For our reaction, the termination steps are.

$$2\ Cl\cdot \rightleftharpoons Cl_2$$

$$Cl\cdot + \cdot CH_2CH_3 \rightleftharpoons CH_2ClCH_3$$

$$2\ \cdot CH_2CH_3 \rightleftharpoons CH_3CH_2CH_2CH_3$$

As you can see from this example, the first termination step generates chlorine, which can break up again via the initiation step, while the second forms useful products. However, the third termination step forms an undesired product; as it turns out, many organic reactions form varying quantities of undesired products, depending on the process that takes place.

Oxidation

Organic compounds undergo a wide variety of oxidation reactions. For example, the oxidation of a primary alcohol results in the formation of an aldehyde, which can be further oxidized to form a carboxylic acid.

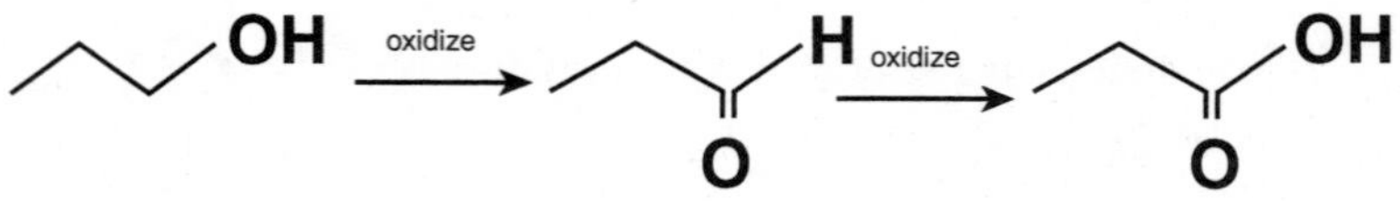

Figure 25.17

The oxidation of 1-propanol results in the formation of propanal. Further oxidation results in the formation of propanoic acid.

Likewise, the oxidation of a secondary alcohol results in the formation of a ketone.

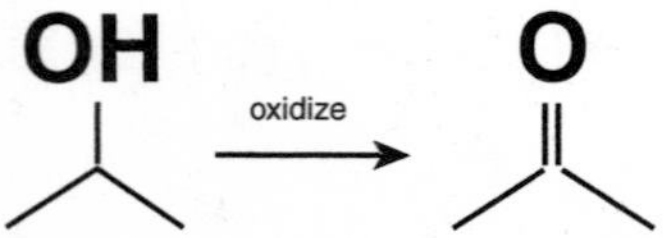

Figure 25.18

When 2-propanol is oxidized, propanone is formed.

Condensation Reactions

In condensation reactions, two molecules combine with each other in a way that results in the formation of water. One example of a condensation reaction is when two molecules of methanol combine with one another to form dimethyl ether and water.

$$CH_3OH + HOCH_3 \longrightarrow H_2O + CH_3\text{-}O\text{-}CH_3$$

Figure 25.19

An example of a condensation reaction. Two molecules of methanol combine with one another to form dimethyl ether and water.

Polymerization Reactions

Back in the beginning of this chapter, I mentioned that carbon was good at forming long chains. One of the ways that this process occurs is by polymerization.

In a polymerization reaction, very small molecules called monomers link up with one another to form much longer chains of molecules called polymers. Most of the plastics you're familiar with are polymers, such as Teflon (polytetrafluoroethene), polyethylene, and polystyrene.

Chemistrivia

Teflon (found on nonstick pots and pans, among other things) was discovered by accident when DuPont scientists, working on new CFC refrigerants, found that they couldn't get the tetrafluoroethylene in a pressurized tank to come out. When the scientists sawed the tanks open, they found that the gas had polymerized and formed an almost completely unreactive solid.

The free-radical reaction that forms polyethylene polymer from ethylene monomers is shown in Figure 25.20.

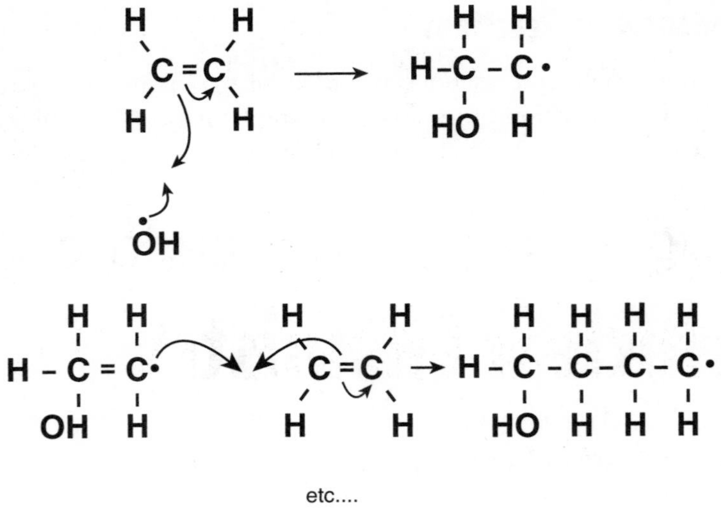

Figure 25.20

The free-radical reaction that forms polyethylene polymer from ethylene monomers.

etc....

When two polymer radicals combine with one another, the chain stops growing. Chemical companies spend considerable time and expense devising reaction conditions that will maximize the lengths of the polymer chains while maintaining high quality and good yields.

The Least You Need to Know

- Hydrocarbons are molecules that contain only carbon and hydrogen. Saturated hydrocarbons (alkanes) have carbon-carbon single bonds, while unsaturated hydrocarbons (alkenes and alkynes) have carbon-carbon multiple bonds.
- Isomers are different molecules that have the same molecular formula.
- There are a whole bunch of functional groups in organic molecules, each of which has its own naming scheme.
- There are many types of reactions involving organic molecules, including addition reactions, substitution reactions, oxidations, condensations, and polymerizations.

Chapter 26

Nuclear Chemistry

In This Chapter

- Commonly used terms
- Types of radioactive decay
- Half-lives
- Binding energy
- Fusion and fission

Nuclear reactions have had an interesting history. In the beginning of the nuclear age, the U.S. government used nuclear weapons to end the second world war and claimed that nuclear power would make electricity so cheap that it would no longer even be metered. Later in the nuclear age, "rogue states" are busy trying to make nuclear weapons, and nuclear power is synonymous in many circles with the term "meltdown."

Unfortunately, most people don't really understand anything at all about nuclear reactions. Maybe I'm a little slow, but it seems to me that to have a strong opinion one way or the other, you need to actually understand something about nuclear reactions. Though I won't take sides in the nuclear debate in this book (I don't really want the hate mail), I hope this chapter gives you the background information you need to make an informed decision one way or the other.

What Are Nuclear Processes?

Nuclear reactions are reactions that involve the nucleus of an atom. There are many types of nuclear reaction, but one thing that they all have in common is that the atom itself is changed, not just the fashion in which it combines with other atoms. As we'll see in this chapter, each type of nuclear reaction has its own characteristics.

Before we can really start to understand nuclear reactions, we need some basic vocabulary.

- Nucleons are the protons and neutrons in an atom. They're referred to as nucleons because they reside in the nucleus.
- Isotopes (discussed extensively in Chapter 4) refer to elements that have the same atomic number but different numbers of neutrons (and thus, different atomic masses). The different isotopes of an element are referred to as nuclides.
- Radioisotopes refer to radioactive nuclear isotopes. Radioactive decay is when a nucleus gives off various small particles. The term "radiation" refers to these small particles released during radioactive decay.

Chemistrivia

Radiation is a term that's often misused. "Electromagnetic radiation" refers to wave phenomena such as light or radio waves—for example, every time you turn on your bedroom light, you're filling the room with electromagnetic radiation. "Ionizing radiation," on the other hand, is the type of radiation given off by radioactive decay. Although we may say that we're going to "nuke a burrito" when we put it in the microwave, the microwave radiation actually has nothing to do with radioactive decay.

- In this chapter, we'll be using the $^{A}_{Z}X$ terminology for denoting nuclides, where A is the atomic mass of the nuclide, Z is the atomic number, and X is the atomic symbol.

Now that we have some very basic terms, let's start talking about nuclear reactions!

Why Does Radioactive Decay Occur?

I want to be honest with you—nobody really knows why radioactive decay occurs. Presumably, there's some physical process that dictates why a nucleus will undergo radioactive decay, but nobody knows what it is. Even though we don't understand why radioactive decay takes place, scientists have come up with a set of rules that seem to do a good job of describing how likely it is that a particular atom will be radioactive.

1. Atoms with more than 83 protons undergo radioactive decay.
2. Nuclides with small masses appear to have roughly a 1:1 ratio of neutrons to protons. For example, ^{12}C has six protons and six neutrons. As the masses of the nuclides increase, the ratio of neutrons to protons also increases, becoming roughly 1.5:1; ^{209}Bi has 126 neutrons and 83 protons, giving it a 1.52:1 neutron/proton ratio. We can show the stable nuclides of each element in a narrow "zone of stability" that has an increasing neutron to proton ratio as the atomic number increases:

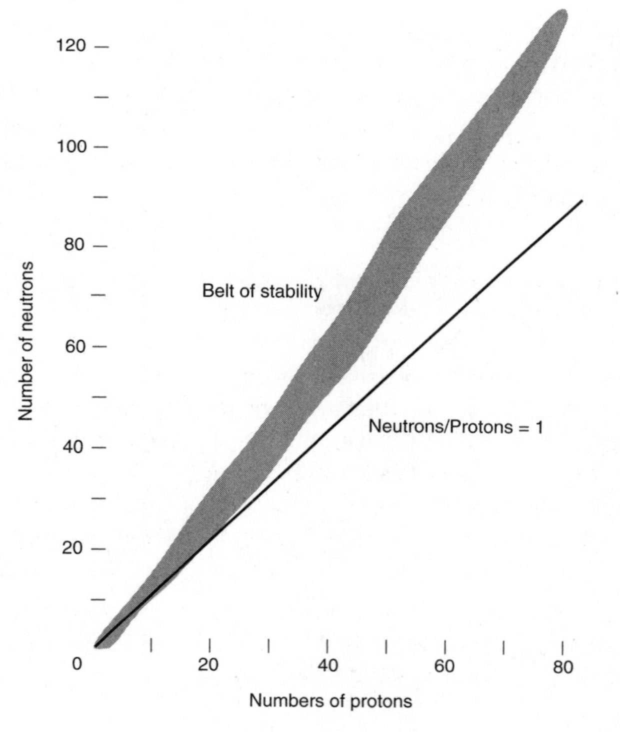

Figure 26.1

Stable nuclides appear to have a very narrow range of neutron/proton ratios.

3. Nuclides containing "magic numbers" of protons or neutrons appear more stable than those with other nuclides. These "magic numbers" are 2, 8, 20, 50, 82, and 126. For example, there are more stable nuclides with 20 protons than stable nuclides with either 19 or 21 protons. Why this is, nobody knows.
4. Nuclides with even numbers of both protons and neutrons are more stable than those with odd numbers of protons and neutrons. Again, nobody knows why this is, but it seems to work.

There are exceptions to these rules, which isn't really surprising, considering that they aren't based on any physical reality that we yet understand. However, if you're not sure if something will be radioactive, these rules are more accurate than random guesses!

Types of Radioactive Decay

There are many ways that atoms can undergo radioactive decay. We'll discuss the possible methods of radioactive decay, as well as how you can make a rough prediction of what type of decay a certain nuclide will undergo.

Alpha Decay

Alpha particles (frequently depicted as the lowercase Greek letter alpha, α) consist of helium nuclei, have the formula ${}_{2}^{4}He$, and a charge of +2. When an element undergoes alpha decay, an alpha particle is given off. Alpha emission causes the atomic number of the element undergoing decay to decrease by two and the atomic mass to decrease by four. An example of alpha decay is shown here.

$${}_{97}^{247}Bk \rightarrow {}_{2}^{4}He + {}_{95}^{243}Am$$

The Mole Says

The equation shown here provides a demonstration of the law of conservation of mass. Notice that the atomic mass of Bk is 247 on the left side of the equation, while the sum of the masses of the alpha particle and Am are also 247. This will be true for all nuclear processes. Likewise, the amount of nuclear charge is 97 for Bk, and the sum of the nuclear charges for He and Am are also 97. This will come in handy later when trying to determine the products of different radioactive processes.

Alpha decay occurs mostly among nuclides with very large masses. For example, 12 of the 18 known isotopes of uranium undergo alpha decay.

Beta Decay

Beta decay, denoted by the Greek letter β, occurs when beta particles (electrons, with a formula of ${}_{-1}^{0}e$) are emitted from the nucleus of an atom. This process effectively converts a neutron to a proton, increasing the atomic number by one without changing the atomic mass. An example of a beta decay is shown here.

$${}_{39}^{102}Y \rightarrow {}_{-1}^{0}e + {}_{40}^{102}Zr$$

Nuclides that undergo beta decay typically have very high neutron to proton ratios and lose beta particles to decrease this ratio. For example, the neutron to proton ratio for ^{102}Y above is 1.62:1.

Gamma Decay

Gamma decay occurs when very high energy light is released from the nucleus of an atom. Called "gamma rays," this very high energy light is usually released during other forms of radioactive decay. Gamma rays, denoted by the Greek letter gamma, γ, appear in nuclear equations as $^{0}_{0}\gamma$ and are electrically neutral.

Positron Emission

Positrons are the antiparticles of electrons and have the symbol $^{0}_{+1}e$. That is to say, positrons have a +1 charge instead of a –1 charge and are instantly annihilated upon contact with an electron. Positron emission results in the conversion of a proton to a neutron, decreasing the atomic number by one but leaving the atomic mass unchanged.

$$^{17}_{9}F \rightarrow ^{17}_{8}O + ^{0}_{+1}e$$

Positron emission most often occurs when an element has a small neutron to proton ratio. In the preceding example, the neutron to proton ratio of fluorine is 0.89:1.

Electron Capture

Electron capture occurs when an electron in an inner orbital is pulled into the nucleus, converting a proton to a neutron. As in positron emission processes, this most often occurs when the neutron to proton ratio of a nuclide is too small. An example is shown here.

$$^{7}_{4}Be + ^{0}_{-1}e \rightarrow ^{7}_{3}Li$$

You've Got Problems

Problem 1: Write the equations for the following decay processes.

(a) Silver-108 undergoes beta decay.

(b) Radon-216 undergoes alpha decay.

Half-Lives

Back in Chapter 21, we discussed half-lives. To recap, the half-life of a reaction is the amount of time it takes for half of the reactant to be converted into products.

In nuclear processes, the concept of half-life is the same as in chemical processes. The half-life in a nuclear reaction is the amount of time it takes for half of the radionuclide atoms to undergo radioactive decay. Fortunately for us, nuclear half-lives use the same rate laws as first-order chemical processes. As a result, the half-life of a nuclear reaction is determined by the following equation:

$$t_{1/2} = \frac{0.693}{k}$$

$t_{1/2}$ is the half-life of the process and k is the rate constant for the nuclear decay process.

Chemistrivia

One of the most important uses of half-lives is carbon dating. Living creatures absorb carbon (including naturally forming radioactive ^{14}C in the food they eat, incorporating it into their tissues. When they die, the non-radioactive ^{12}C remains, while the radioactive ^{14}C, with a half-life of 5,730 years, slowly vanishes through beta decay. By comparing the quantity of ^{14}C in a sample to the quantity of ^{14}C in living creatures, the ages of formerly living objects can be accurately determined.

Let's see a sample problem of a half-life calculation:

Example: Determine the following for the alpha decay of ^{236}Pu:

1. What is the rate constant for this process, given that the half-life is 87.74 years?
2. If I have 175 grams of ^{236}Pu, how many grams will be left after 225 years?

Solution:

1. Determining the rate constant is fairly simple because all we need to do is plug the value for half-life into the equation we were given:

$$t_{1/2} = \frac{0.693}{k}$$

$$k = \frac{0.693}{t_{1/2}} = \frac{0.693}{87.74 yrs} = 7.90x10^{-3} / yr$$

2. To determine how much ^{236}Pu will be left over, we'll need to use the rate constant as well as the equation for determining the relationship between reactant and time for a first order process (Chapter 21).

$$\ln[A]_t = -kt + \ln[A_o]$$

For this equation, $[A_o]$ is the initial quantity of the reactant, $[A]_t$ is the quantity of the reactant at time = $t_{[1/2]}$, k is the rate constant, and $t_{[1/2]}$ is the half-life. Placing our values into this equation, we get:

$$\ln[A]_t = -(7.90 \times 10^{-3}/\text{yr})(225\text{ yr}) + \ln[175\text{ g}]$$

$$\ln[A]_t = -1.78 + 5.16$$

$$[A]_t = 29.4\text{ grams}$$

You've Got Problems

Problem 2: Determine the rate constant for the alpha decay of ^{148}Gd, given that the half life of this process is 75 years.

Binding Energy—Relating Mass to Energy

Let's do some simple math. The mass of a proton is 1.00728 amu. The mass of a neutron is 1.00867 amu. If I have 92 protons and 146 neutrons in a nucleus, the nucleus should weigh:

$$\text{Weight} = (92 \times \text{mass of proton}) + (146 \times \text{mass of neutron})$$

$$= (92 \times 1.00728\text{ amu}) + (146 \times 1.00867\text{ amu})$$

$$= 239.9356\text{ amu}$$

The nucleus that has 92 protons and 146 neutrons is the ^{238}U nuclide. However, the nucleus of a ^{238}U atom actually weighs 238.0003 amu. What happened to the other 1.9353 amu?

Chemistrivia

The "missing mass" in a nucleus is referred to as the "mass defect."

As it turns out, the missing mass has been converted to energy. Albert Einstein said that mass and energy could be converted to one another by the equation $E = mc^2$, where E is energy, m is mass, and c is the speed of light (3.00×10^8 m/s). In our preceding example, the missing mass is in the form of energy that holds the nucleus together.

To figure out how much energy there really is for each atom, let's solve the equation $E = mc^2$ for the information we were given. Because 1.00 amu = 1.67×10^{-27} kg, we will convert amu to kilogram by multiplying 1.9353 amu by 1.67×10^{-27} kg to get a mass of 3.23×10^{-27} kg.

$$E = mc^2$$

$$= (3.23 \times 10^{-27} \text{ kg})(3.00 \times 10^8 \text{ m/s})^2$$

$$= 2.91 \times 10^{-10} \text{ J}$$

The unit "J" stands for "Joules" and is equal to 1 kg m^2/s^2.

Now, this may not seem like very much energy. However, remember that this is only for one nucleus of uranium. In every mole of uranium, there are 6.02×10^{23} nuclei, with a volume of less than 13 milliliters. However, the binding energy in one mole of uranium is 1.75×10^{14} J, enough energy to raise the temperature of 465 million kilograms of water by 90° C. That's a lot of energy!

Chemistrivia

Nuclear power plants use nuclear binding energies to generate large quantities of heat. This heat is used to boil water, and the steam causes electrical turbines to generate electricity. The only difference between a nuclear power plant and one that uses fossil fuels such as oil or coal is that the source of heat to boil the water is different—both types of power plant use steam to turn turbines.

Nuclear Fission

Nuclear fission is one of the processes that allows us to generate large quantities of energy from nuclear processes. In a *fission reaction*, a heavy nucleus is hit with a neutron, which causes it to break apart into smaller elements. Because some of the energy that was used to hold the nucleus together is turned back into mass, the process gives off a huge amount of heat.

def•i•ni•tion

Fission reactions are nuclear processes in which large nuclei are split apart to produce smaller nuclei and very large amounts of energy.

One of the radioisotopes used in fission reactions is ^{235}U. When a neutron (^1_0n) hits an atom of ^{235}U, the following processes occur.

$$^1_0n + ^{235}_{92}U \rightarrow ^{137}_{52}Te + ^{97}_{40}Zr + 2^1_0n$$

$$^1_0n + ^{235}_{92}U \rightarrow ^{142}_{56}Ba + ^{91}_{36}Kr + 3^1_0n$$

When either of these processes occurs, you can see that more neutrons are formed than were required to make the uranium-235 nucleus break apart. As this process continues, more and more neutrons are generated, each of which splits another uranium nucleus. Because this process is self-perpetuating and repeats itself many times, it's said to be a chain reaction. A picture of what this looks like is shown in Figure 26.2.

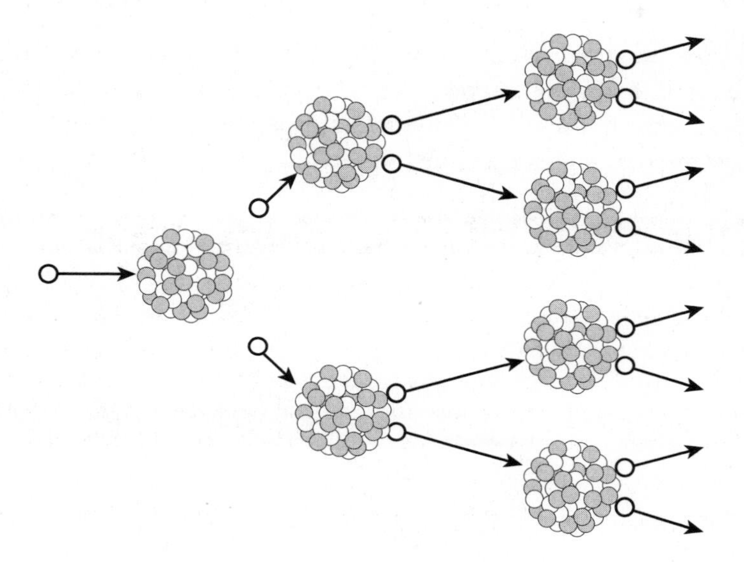

Figure 26.2

When one uranium atom is broken apart, the neutrons that are formed cause neighboring uranium atoms to also break apart.

However, it's not enough to fire a neutron at any piece of uranium, because if not enough uranium is present, the neutrons will escape more often than they split other uranium nuclei apart. It's only when we get a large enough quantity of uranium (called a "critical mass") that the reaction proceeds at a constant rate. For the reaction to proceed at an expanding rate, an even larger amount of uranium (called a "supercritical mass") is required.

Fusion Reactions

Fusion reactions occur when small nuclei are stuck together to form larger nuclei. Fusion reactions generate heat because the binding energy that holds the nuclei together is less in the products than in the initial nuclides—as a result, the additional energy is given off as heat.

def•i•ni•tion

Fusion reactions are nuclear processes in which small nuclei combine to make larger ones plus a huge quantity of energy.

Fusion reactions are difficult to start because it takes a huge amount of energy to make them occur. The fusion reaction that requires the least amount of energy is the combination of deuterium (hydrogen-2) with tritium (hydrogen-3) to form helium-4 and a neutron.

$$^{2}_{1}H + ^{3}_{1}H \rightarrow ^{4}_{2}He + ^{1}_{0}n$$

This process requires temperatures of approximately 40,000,000 K to occur. Because so much heat is required to make this happen, the only practical use of nuclear fusion has been in nuclear weapons. In these "thermonuclear" weapons or "hydrogen bombs," the fusion reaction is triggered by the intense heat of a fission reaction.

Chemistrivia

Though there is hope that fusion reactions will someday be able to generate huge quantities of energy, it's unlikely that this will happen anytime soon. Currently, it takes far more energy to make a small fusion reaction occur than is generated by the reaction itself. In 1989, the scientists Pons and Fleischmann generated a great deal of publicity with "cold fusion," which purported to create fusion reactions at room temperature via catalysis. Most scientists believe that cold fusion is nonsense, though a vocal minority continue with their research.

The Least You Need to Know

- Radioactive decay occurs because some atomic nuclei are, for reasons not yet understood, unstable.
- The main types of radiation are alpha decay, beta decay, gamma decay, positron emission, and electron capture.
- The amount of time it takes for half of a radioactive sample to decay is its half-life.
- The amount of mass that's converted to energy in a nucleus is called the "binding energy."
- Fission is when a nucleus breaks apart to form smaller nuclei, while fusion is when smaller atoms join to form a larger nucleus.

Part 6 Thermodynamics 101

In my chemistry classes, I like to do demonstrations where things explode or catch fire. After all, who doesn't like to see something blow up?

As it turns out, things don't just blow up for no reason (though it may sometimes seem that way). They blow up because energy, like anything else, behaves according to certain rules that take some getting used to. These rules are called thermodynamics.

In this part, we'll learn why some reactions occur spontaneously, while others don't proceed at all. Of course, this involves learning fancy terms like "entropy," "enthalpy," and "free energy," but you'll get the hang of it in no time!

Chapter 27

Cranking Up the Heat: Basic Thermodynamics

In This Chapter

- Energy, heat, and temperature
- Enthalpies of formation
- Finding enthalpies of reaction using Hess's Law
- Calorimetry

As we approach the end of this book, it's clear that we haven't spent much time on energy. Sure, I mentioned it a little bit in Chapter 20 when I told you about energy diagrams, but we haven't really talked about what energy is or how it works.

Well, that's about to change. Pull out your calculators and let's get ready to crank up the heat!

What Is Energy?

This seems like a simple question. After all, if you put your hands on a hot stove burner, you know from your flaming fingers that energy has been transferred from the burner to your hands. However, having an intuitive feel for what energy is isn't the same thing as defining it.

Energy is defined as the capacity of something to do work or produce heat. For example, if I feed my very young nephew 15 chocolate bars, I'll find out from his hyperactive behavior that the chocolate contained a considerable quantity of energy.

There are two kinds of energy that we need to concern ourselves with.

- Kinetic energy is energy having to do with how fast something is moving. For example, if my chocolate-filled nephew threw a soccer ball at your head, you'd certainly realize that the soccer ball contained kinetic energy. Kinetic energy can be calculated by the equation $KE = \frac{1}{2}mv^2$, where m is the mass of the object in kg and v is the velocity of the object in m/s.
- Potential energy is stored energy. For example, if I pour two liters of gasoline on my neighbor's van when the alarm goes off at 4 A.M., nothing happens. However, if I throw a match on the van, we'll see a very fast release of energy, not to mention a very angry neighbor.

def•i•ni•tion

Energy is the capacity of an object to do work or produce heat. Kinetic energy is described as the energy associated with motion, and potential energy is stored energy.
The **law of conservation of energy** (also called the first law of thermodynamics) states that energy can neither be created nor destroyed in any process.

As it turns out, the amount of energy in the universe is constant, though it frequently changes form. The idea that energy can be neither created nor destroyed is called the "first law of thermodynamics," or the *law of conservation of energy*. As a result, energy can change form, but can never be destroyed or created. For example, the kinetic energy created when I set my neighbor's van ablaze is equal to the potential energy stored in the chemical bonds of the gasoline.

Chemistrivia

Food energy content is commonly given in "Calories." The capital "C" at the front of this unit makes it different from the other "calorie" we discussed. 1 Calorie (food) = 1 kilocalorie.

The unit used to describe energy is the "joule"— $1 J = 1 kg \cdot m^2/s^2$. Another common unit of energy is the "calorie," which is defined as the amount of energy required to heat one gram of water by 1° C. There are 4.184 J in 1 cal.

Temperature and Heat

Temperature and heat are not the same thing. This contradicts the way you probably see the world. For example, if somebody asks you how hot it is outside, you'd probably say that the temperature was 25° C. This may be true, but it doesn't answer the question.

The term *temperature* describes the amount of motion that the molecules or atoms in a material have. If these particles are moving very quickly, the material has a high temperature. If the particles move slowly, the material has a low temperature.

Heat, on the other hand, describes the amount of energy that is transferred from one object to another. When you go outside on a warm day, you don't feel hot because the air molecules are moving quickly—you feel hot because these molecules collide and transfer some of their energy to your skin. Though this is a subtle difference, it will become important later. The study of heat is called "thermodynamics."

def•i•ni•tion

Temperature describes the motions of the particles in a material, and **heat** describes the amount of energy moved from one object to another.

Describing Energy Changes

Let's say that I have a closed can of beans that has been sitting in the trunk of my car on a hot day, bringing its temperature to 50° C. Now, imagine that I put this can of beans into a bucket of ice. As you can probably guess, the energy of the beans will be transferred to the ice, causing it to melt.

In thermodynamic terms, we would think of the can of beans as being the "system" and the ice as being the "surroundings." Using this terminology, we would say that the can of beans lost energy because it was transferred to its surroundings. As a result, the change in energy for the beans would be negative. The concept of "change" is shown by the symbol "Δ," so the change in energy for the beans is ΔE.

On the other hand, if we placed the hot beans into a campfire, energy would be transferred from the campfire to the beans. As a result, the change in energy for the beans would be positive ($\Delta E > 0$).

The change in energy, ΔE, for a process can be said to be equal to the difference between the energy of the system after the process and the energy of the system before the process. In equation form:

$$\Delta E = \Delta E_{final} - \Delta E_{initial}$$

Energy is defined as the capacity of a system to produce heat or do work, so we can write this equation another way. To restate, the energy change for a process is equal to the change in heat for the process ("q") and the amount of work performed by the system for the process ("w").

$$\Delta E = q + w$$

If the change in heat for a process is positive, the process is said to be "endothermic" because the process won't occur without the addition of energy. If the change in heat for a process is negative, the process is said to be "exothermic" because the process causes the system to release excess heat into its surroundings. Generally, processes that feel cold (such as the reaction that takes place in a chemical cold pack) are endothermic while processes that feel hot (such as setting my neighbor's van on fire) are exothermic.

Energy Is a State Function

Upon reading the above statement, you're probably asking yourself, "What the heck is a state function?" This means that the energy of a system depends on the conditions present in the material, such as the temperature, pressure, and quantity of the material. It doesn't matter where the material came from—the only thing that matters is its current condition.

P-V Work

One type of work that's fairly common in chemical processes is work having to do with the expansion or contraction of gases. In the example of an automobile, the expanding gases in the cylinders cause a piston to move, which ultimately causes the car to move forward.

Figure 27.1 shows how a gas performs work in a piston.

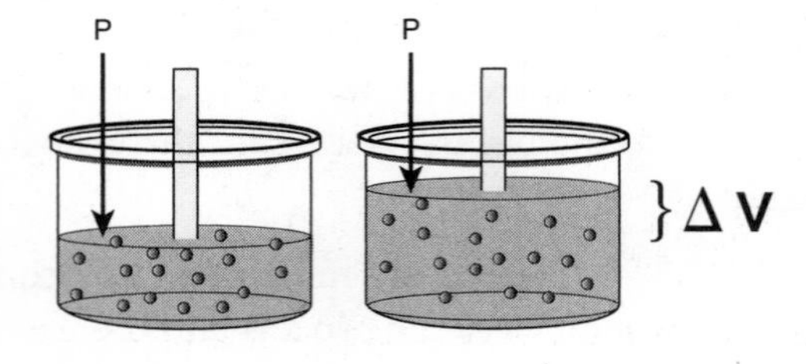

Figure 27.1

When the gas in a cylinder expands, the product of the change in volume (ΔV) and the outside pressure (P) is equal to the amount of work performed.

A gas in a cylinder performs work by pushing on a piston. When the piston is pushed outward, the difference in the initial and final volume inside the cylinder is called ΔV. Because there is external atmospheric pressure exerted on the piston (P), the amount of work the gas performs will be:

$$w = -P\Delta V$$

The sign for work in this expression is negative because the gas is performing work on its environment. Earlier, we defined the change in energy for a process as:

$$\Delta E = q + w$$

Thus, we can substitute $P\Delta V$ for w to get the equation:

$$\Delta E = q_P - P\Delta V$$

The heat term is described as q_P and not just q to express the idea that this system is operating under constant external pressure. By moving the terms in this equation around, we can find that:

$$q_P = \Delta E + P\Delta V$$

The change in heat under constant pressure is called enthalpy. *Enthalpy* has the symbol H and is defined as the amount of heat that flows to or from a system at constant pressure. As a result, the change in enthalpy for a process is defined as:

$$\Delta H = \Delta E + P\Delta V$$

def•i•ni•tion

Enthalpy (H) is the amount of heat in a system at constant pressure.

Enthalpy, like energy, is a state function, so the amount of enthalpy for a process depends on the quantity of material that is undergoing a change. For example, if the enthalpy change produced by burning one mole of a substance is –150 kJ, the amount of energy change produced by burning two moles of a substance will be twice that, or –300 kJ.

Enthalpies of Chemical Reactions

The enthalpy change for a chemical reaction is equal to the sum of the enthalpies of the products minus the sum of the enthalpies of the reactants:

$$\Delta H = H_{products} - H_{reactants}$$

def•i•ni•tion

The enthalpy change for a chemical reaction is more commonly known by the term **heat of reaction** and has the symbol ΔH_{rxn}. The enthalpy required for a chemical to be created from its elements is called the **heat of formation** and has the symbol ΔH_f.

One of the main ways this comes in handy is when we try to determine the enthalpy change of a chemical reaction given the enthalpy required to form the reactants and products.

When the *heats of formation* for a series of chemical compounds are given, the symbol is usually given as ΔH^o_f. The little "o" after the ΔH term appears insignificant, but means that the heat of formation given is for the form of the substance that's most stable at a pressure of 1 atm and 298 K (25° C). The proper term for a heat of formation under these conditions is "standard heat of formation" because it's determined at standard conditions.

The Mole Says

The standard heats of formation for pure elements is said to be zero, as long as the elements are in their most stable form. For example, the standard heat of formation for O_2 is zero, but the standard heat of formation for O_3 is higher because it's less stable than O_2 under standard conditions.

To understand what all that stuff I just wrote really means, let's do an example.

Example: Find the heat of combustion of ethene given the following information.

$\Delta H^o_f(C_2H_{4(g)}) = +52.3$ kJ/mol

$\Delta H^o_f(CO_{2(g)}) = -393.5$ kJ/mol

$\Delta H^o_f(H_2O_{(l)}) = -285.8$ kJ/mol

Solution: Before we can do anything, we must write a balanced equation for the combustion of ethene.

$$C_2H_{4(g)} + 3\ O_{2(g)} \leftrightarrow 2\ CO_{2(g)} + 2\ H_2O_{(l)}$$

The heat of combustion for this process will be equal to the sums of the heats of formation for the products minus the sums of the heats of formation for the reactants.

Products:

ΔH^o_f for 2 CO_2 = 2 mol × –393.5 kJ/mol = –787.0 kJ

ΔH^o_f for 2 H_2O = 2 mol × –285.8 kJ/mol = –571.6 kJ

Total: –1358.6 kJ

Reactants:

ΔH^o_f for 1 C_2H_4 = 1 mol × 52.3 kJ/mol = 52.3 kJ

ΔH^o_f for 2 O_2 = 2 mol × 0.00 kJ/mol = 0.00 kJ

Total: +52.3 kJ

Thus:

$\Delta H^o_{rxn} = \Delta H^o_f\text{ (products)} - \Delta H^o_f\text{(reactants)}$

ΔH^o_{rxn} = –1358.6 kJ – 52.3 kJ

ΔH^o_{rxn} = –1410.9 kJ

You've Got Problems

Problem 1: Determine the heat of combustion of sucrose ($C_{12}H_{22}O_{11(s)}$) given the following standard heats of formation:

ΔH^o_f ($C_{12}H_{22}O_{11(s)}$) = −2221 kJ/mol

ΔH^o_f ($CO_{2(g)}$) = −393.5 kJ/mol

ΔH^o_f ($H_2O_{(l)}$) = −285.8 kJ/mol

Hess's Law

Sometimes, the reactants in a chemical process need to undergo several changes to become products. For such processes, the sums of the enthalpy changes for all of the steps are equal to the overall enthalpy for the process. This is known as Hess's Law.

Let's use Hess's Law to determine the enthalpy of formation of N_2O_5. The formation reaction is given by the following equation.

$$2\ N_{2(g)} + 5\ O_{2(g)} \leftrightarrow 2\ N_2O_{5(g)}$$

We don't have any tables including heats of formation handy. Fortunately, we do have the following standard heats of reaction.

$$2\ NO_{(g)} + O_{2(g)} \leftrightarrow 2\ NO_{2(g)}\ \Delta H^o_{rxn} = -114\text{ kJ/mol}$$

$$4\ NO_{2(g)} + O_{2(g)} \leftrightarrow 2\ N_2O_{5(g)}\ \Delta H^o_{rxn} = -110\text{ kJ/mol}$$

$$N_{2(g)} + O_{2(g)} \leftrightarrow 2\ NO_{(g)}\ \Delta H^o_{rxn} = +181\text{ kJ/mol}$$

You may be wondering why it's fortunate that we have this information. After all, what do all of these reactions have to do with the one we're interested in?

I'm sure glad you asked! By combining the preceding equations, we can come up with a whole new equation that describes the process we're actually interested in. The only rules we need to follow are these:

- If we reverse a reaction, the sign of the standard heat of reaction is reversed. For example, the process:

  $2\ NO_{2(g)} \leftrightarrow 2\ NO_{(g)} + O_{2(g)}$

  has a $\Delta H^o_{rxn} = +\ 114$ kJ/mol
- If we need to perform a reaction more than once, the heat of reaction is multiplied by the number of times we do the reaction. For example, if we do the reaction:

  $2\ NO_{2(g)} \leftrightarrow 2\ NO_{(g)} + O_{2(g)}$

  twice, the heat of reaction will be 2 mol × 114 kJ/mol = 228 kJ.

Let's use Hess's Law to determine the heat of formation of $N_2O_{5(g)}$.

$$2\ N_{2(g)} + 5\ O_{2(g)} \leftrightarrow 2\ N_2O_{5(g)}$$

Step 1

There are two moles of N_2 in the reactants. The only equation that contains nitrogen is:

$$N_{2(g)} + O_{2(g)} \leftrightarrow 2\ NO_{(g)}\ \ \Delta H^o_{rxn} = +\ 181 \text{ kJ/mol}$$

so we'll multiply this equation by two to give us

$$2\ N_{2(g)} + 2\ O_{2(g)} \leftrightarrow 4\ NO_{(g)}\ \ \Delta H^o_{rxn} = +362 \text{ kJ}$$

Step 2

The equation we're trying to solve has two moles of $N_2O_{5(g)}$ as the products. The only equation that contains $N_2O_{5(g)}$ is:

$$4\ NO_{2(g)} + O_{2(g)} \leftrightarrow 2\ N_2O_{5(g)\ \Delta}H^o_{rxn} = -110 \text{ kJ}$$

Because there are already two moles of $N_2O_{5(g)}$ in the products, we can leave the heat of reaction the way it is.

Step 3

Now that we have two equations, let's add them together to see what we still need to do.

$$2\ N_{2(g)} + 2\ O_{2(g)} \rightleftharpoons 4\ NO_{(g)}\ \ \Delta H^\circ_{rxn} = +362\ kJ$$

$$4\ NO_{2(g)} + O_{2(g)} \rightleftharpoons 2\ N_2O_{5(g)}\ \ \Delta H^\circ_{rxn} = -110\ kJ$$

Overall:

$$2\ N_{2(g)} + 3\ O_{2(g)} + 4\ NO_{2(g)} \rightleftharpoons 4\ NO_{(g)} + 2\ N_2O_{5(g)}$$

$$\Delta H^\circ_{rxn} = +252\ kJ$$

Step 4

Somehow, we need to get rid of the $NO_{2(g)}$ on the reactant side of the equation and the $NO_{(g)}$ on the product side. Fortunately, we have an equation to help us do this.

$$2\ NO_{(g)} + O_{2(g)} \rightleftharpoons 2\ NO_{2(g)}\ \ \Delta H^\circ_{rxn} = -114\ kJ/mol$$

By multiplying this equation by 2, the $NO_{(g)}$ on the product side and the $NO_{2(g)}$ on the reactant side cancel out.

$$4\ NO_{(g)} + 2\ O_{2(g)} \rightleftharpoons 4\ NO_{2(g)}\ \ \Delta H^\circ_{rxn} = -228\ kJ$$

Step 5

Adding this equation to the others, we get:

$$2\ N_{2(g)} + 3\ O_{2(g)} + \cancel{4\ NO_{2(g)}} + \cancel{4\ NO_{(g)}} + 2\ O_{2(g)} \rightleftharpoons 4\ NO_{(g)} + 2\ N_2O_{5(g)} + 4\ NO_{2(g)}$$

or

$$2\ N_{2(g)} + 5\ O_{2(g)} \rightleftharpoons 2\ N_2O_{5(g)}$$

By adding the heats of reaction, we can find that the total standard heat of formation for this process is:

$$\Delta H^\circ_f = +252\ kJ - 228\ kJ = 24\ kJ$$

Problems such as this one take time and practice. However, with a bit of trial and error, you should be able to figure out the heat of reaction for just about any process.

You've Got Problems

Problem 2: Find the heat of reaction for the following process.

C (graphite) $\rightleftharpoons$ C (diamond)

Given the following information:

C(diamond) + $O_{2(g)} \rightleftharpoons CO_{2(g)}$ $\Delta H_{rxn} = -395.4$ kJ

C(graphite) + $O_{2(g)} \rightleftharpoons CO_{2(g)}$ $\Delta H_{rxn} = -393.5$ kJ

Calorimetry

Calorimetry is a process by which the energy change of a process can be experimentally determined. Calorimetry works by performing a chemical reaction within a steel container called a "bomb" that is immersed in a giant bucket of water. Because the energy that is produced by the chemical reaction is transferred to the water, the temperature change of the water can be used to determine the heat of reaction.

A bomb calorimeter is shown in Figure 27.2.

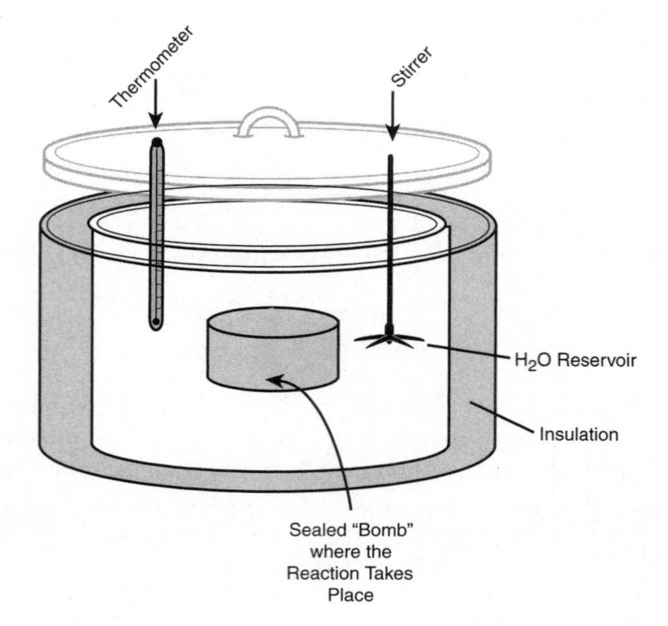

Figure 27.2
A bomb calorimeter.

The amount of energy required to raise the temperature of a substance by 1° C is called the heat capacity (C_p). This is useful in calorimetry because by knowing how much water is in the calorimeter and the heat capacity of water (4.184 J/g° C), the amount of heat generated can be calculated.

The equation used to do this is shown below.

$$\Delta H = mC_p\Delta T$$

Where ΔH is the amount of energy transferred from the bomb to the water, m is the mass of water (in g), C_p is the heat capacity of water (4.184 J/g° C), and ΔT is the change in water temperature (in °C).

Let's see an example of a calculation involving bomb calorimetry.

Example: I have set up a calorimetry experiment in which I will burn 5.00 grams of anthracene ($C_{14}H_{10}$) in a bomb surrounded by 2.00 kg of water. If the temperature of the water rose by 24.1° C, what is the molar heat of combustion of anthracene?

Solution: The first step is to figure out how much energy was released by this process. Since we have 2,000 g of water, its heat capacity is 4.184 J/g° C, and the temperature change is 24.1° C, we can use the following equation to determine how much heat was transferred into the water.

$$\Delta H = mC_p\Delta T$$

$$\Delta H = (2{,}000\ g)(4.184\ J/g°\ C)(24.1°\ C)$$

$$\Delta H = 202{,}000\ J$$

But wait, there's more! We're trying to find the molar heat of combustion of anthracene, but we only burned 5.00 grams. The molar mass of anthracene is 178 g/mol, which means that we burned only 0.0281 mol of anthracene. To determine the molar heat of combustion, we need to figure out how much energy was released for every gram of anthracene burned and multiply this value by the molar mass of anthracene. This calculation is shown here.

$$\frac{202{,}000J}{0.0281mol} = 7{,}190{,}000J / mol$$

This is very close to the actual value for the heat of combustion of anthracene, 7,064 kJ/mol, reflecting some experimental error.

You've Got Problems

Problem 3: I am performing an experiment in which I will burn 1.00 grams of naphthalene ($C_{10}H_8$) in a bomb calorimeter. If the bomb is immersed in a bucket containing 1,500 grams of water and the heat of combustion of naphthalene is 5,154 kJ/mol, how much would you expect the temperature of the water to rise?

The Least You Need to Know

- Kinetic energy is energy of motion, and potential energy is stored energy. Energy can be converted from one form to the other, but can never be created or destroyed (1st law of thermodynamics.)
- Enthalpy is the heat change for a system at constant pressure.
- When given the standard heats of formation for the products and reactants in a chemical reaction, the standard heat of reaction can be found by subtracting the sums of the heats of formation for the products by the sums of the heats of formation for the reactants.
- Hess's Law states that the enthalpy change for a multistep process is equal to the sums of the enthalpy changes for each step.
- Calorimetry is the primary method used to determine the heats of reaction for chemical processes.

Chapter 28

Thermodynamics and Spontaneity

In This Chapter

- Spontaneous processes
- Entropy
- Free energy

You may think that your understanding of enthalpy has made you into a superhero, able to leap tall reactions in a single bound. Well, I hate to break the news to you, but it just ain't so.

You see, there's more to chemical reactions than the simple heat of reaction. As you'll learn in this chapter, heats of reaction play only a part in determining whether or not a reaction will occur spontaneously.

Spontaneous Processes

Here's a pop quiz for you: Which of these processes may occur spontaneously?

- An apple falls upward, attaching itself to the branches of a tree.
- Water trickles down the side of a mountain after a rainstorm.
- Politicians speak honestly and forthrightly about important issues, without lying to make themselves look good.

If you guessed that only the second process may be spontaneous, you understand what spontaneity is. In the world of chemistry, just as in the examples above, spontaneous processes are those that take place without outside intervention.

As it turns out, the reverse of spontaneous processes are not spontaneous. For example, water won't trickle up a mountain before a rainstorm. Likewise, the reverse of non-spontaneous processes are spontaneous—apples fall out of trees and politicians lie like crazy (that is, unless you're a politician, in which case the preceding example doesn't apply to you).

Some Random Thoughts on Entropy

As it turns out, a major driving force for spontaneous processes is something called *entropy*, which has the symbol "S." Entropy is defined as being a measure of the randomness of a system.

You're probably already familiar with the basic idea of entropy. My brother and his wife have a two-year-old son who likes to run around the house and make a mess. Every morning my sister-in-law makes the house spotless, and by the evening, my nephew has turned the house into a disaster area. By his actions (which are, by all accounts, spontaneous), the house becomes more random.

def•i•ni•tion

Entropy (S) is a measure of the randomness of a system.

In a simple example involving chemistry-type stuff, let's imagine two flasks connected with a valve. One of the flasks contains 1 atm of nitrogen gas and the other is in a complete vacuum (Figure 28.1):

Figure 28.1

In the first diagram, all of the nitrogen gas is in the first flask. However, in the second diagram, when the valve is opened, gas rushes into the second flask.

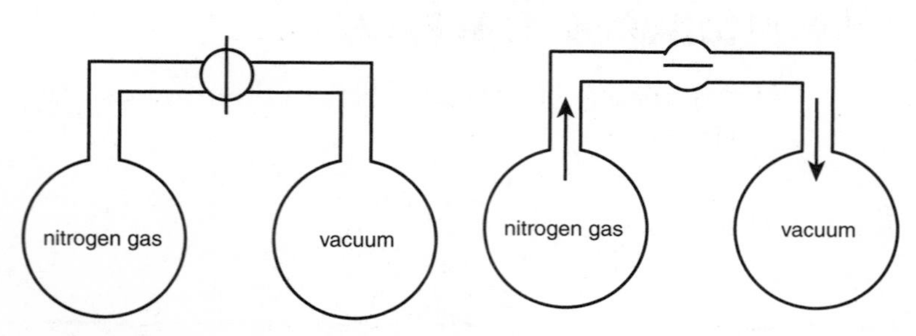

As you can see, the system of flasks is initially ordered with all of the gas in one flask and nothing in the other. However, when we open the valve, the gas will spontaneously flow into the second flask, leaving the system more random than before the valve was opened. The driving force for this process was entropy.

The behavior of entropy is spelled out by *the second law of thermodynamics*. This law states that spontaneous processes are always accompanied by an increase in randomness (entropy) in the universe.

def•i•ni•tion

The second law of thermodynamics states that the entropy is always positive for spontaneous processes.

In math terms (because we all love math!), this is shown as:

$$\Delta S_{universe} = \Delta S_{system} + \Delta S_{surroundings} > 0$$

for spontaneous processes. Note that this equation doesn't say that the randomness of a *system* has to always increase, just that the randomness of the *universe* has to increase.

Bad Reactions

If you plug a vacuum cleaner into the wall and pick up huge quantities of dirt from the floor, your house is made less random. Though this appears to decrease the randomness of the universe, keep in mind that the process used to generate the electricity caused a huge increase in entropy! No process that increases the order in a system can take place unless there's a process that creates an even larger disorder in the surroundings!

It probably isn't a surprise to find that the entropy of a solid is less than that of a liquid, and that the entropy of a liquid is much less than that of a gas. After all, the molecules in a solid are locked tightly in place, the molecules in a liquid are close to one another but can move freely, and the molecules in a gas interact very little with one another, behaving almost as if the other molecules aren't even there. As a result, processes that create liquids from solids or gases from liquids are accompanied by an increase in entropy. To put this in mathematical terms:

$$\Delta S_{solid} < \Delta S_{liquid} << \Delta S_{gas}$$

The Mole Says

The idea that solids have less entropy than other states of matter is spelled out in the third law of thermodynamics, which states that the entropy of a pure crystal at absolute zero (0 K) is zero. As the crystal is warmed, the entropy increases, and each phase change is accompanied by a larger increase in entropy. Incidentially, it's impossible to reach a temperature of exactly 0 K, so everything in the real world has some entropy.

Like enthalpy, entropy is a state function. Entropy depends on the physical conditions of a material, including temperature, pressure, and quantity of material present. It doesn't matter where the material has been or how it was made—its current conditions are sufficient to define its entropy.

Calculating Entropy Changes

In Chapter 27, we mentioned that the enthalpy of a process can be calculated by subtracting the enthalpies of the reactants from the enthalpies of the products. Likewise, we can determine the entropy of a reaction by subtracting the entropies of the reactants from the entropies of the products. In equation form, this is:

$$\Delta S^o_{rxn} = S^o_{products} - S^o_{reactants}$$

Example: Calculate ΔS^o_{rxn} for the reaction:

$$2\ C_{(s)} + 3\ H_{2(g)} \rightleftharpoons C_2H_{6(g)}$$

Given the following information:

$\Delta S^o(C_{(s)}) = 5.7$ J/mol K

$\Delta S^o(H_{2(g)}) = 130.6$ J/mol K

$\Delta S^o(C_2H_{6(g)}) = 229.5$ J/mol K

The Mole Says

Recall that the little "o" above the ΔS term in the example reflects the fact that the reaction takes place under standard conditions: 1 atm pressure, 298 K, 1 M concentration for solutions, and all solids and liquids are in their pure forms.

Solution: To solve this problem, we need to subtract the entropies of the reactants from the entropies of the products.

Reactants:

ΔS^o for 2 mol C = 2 mol × 5.7 J/mol K = 11.4 J/K

ΔS^o for 3 mol H_2 = 3 mol × 130.6 J/mol K = 391.8 J/K

Total: 403.2 J/K

Products:

ΔS° for 1 mol C_2H_6 = 1 mol × 229.5 J/mol K = 229.5 J/K

$\Delta S^\circ_{rxn} = \Delta S^\circ_{products} - \Delta S^\circ_{reactants}$

= 229.5 J/K – 403.2 J/K

=–173.7 J/K

The units for entropy are "J/K."

Our result shows that the entropy change for this process is negative, which isn't surprising considering that the reactants contain three moles of gas and the products consist of only one mole of gas. Because there are fewer moles of gas, we would expect the products to be more ordered than the reactants.

You've Got Problems

Problem 1: Determine ΔS°_{rxn} for the reaction:

$FeCl_{3(s)} + 3\ Na_{(s)} \rightleftharpoons Fe_{(s)} + 3\ NaCl_{(s)}$

Given the following information:

$\Delta S^\circ(FeCl_{3(s)})$ = 142.3 J/mol K

$\Delta S^\circ(Na_{(s)})$ = 51.3 J/mol K

$\Delta S^\circ(Fe_{(s)})$ = 27.2 J/mol K

$\Delta S^\circ(NaCl_{(s)})$ = 72.3 J/mol K.

Free Energy

Now that we know about entropy and enthalpy, we have everything we need to determine, once and for all, if a chemical reaction is spontaneous. *Free energy (G)* is defined as the capacity of a system to do work, and is related to both entropy and enthalpy.

To determine whether a process can be spontaneous, we need to calculate the change in free energy for the process using the following equation.

$\Delta G = \Delta H - T\Delta S$

def•i•ni•tion

Free energy (G) is the capacity of a system to do work. It is also referred to as "Gibbs free energy," which is where the "G" comes from.

The ΔH term in this equation reflects the fact that exothermic processes are more likely to be spontaneous than endothermic processes. The ΔS term reflects the fact that chemical reactions in which the entropy increases are more likely to be spontaneous than those that have decreasing entropy. The temperature term reflects the increased importance of entropy on free energy at higher temperatures.

The free energy change of a process is used directly to compute whether a reaction is spontaneous.

- If ΔG is negative, the reaction is spontaneous.
- If ΔG is zero, the reaction is at equilibrium.
- If ΔG is positive, the forward reaction is not spontaneous, but the reverse reaction is.

Earlier, I told you that processes with a positive ΔS are spontaneous, so you may be wondering why we need ΔG at all. This is true—when ΔS is positive for the entire universe, the process will proceed. However, we rarely have access to the entire universe, so when we calculate whether an isolated chemical reaction is spontaneous or not, we're stuck with using ΔG. This takes into account the entropy change of the system (but not of the rest of the universe), as well as the system's enthalpy change.

Calculating Changes in ΔG°

Just like entropy and enthalpy, we can find the free energy change for a reaction by subtracting the free energies of formation of the reactants from those of the products:

$$\Delta G_{rxn} = G_{products} - G_{reactants}$$

Example: Calculate ΔG_{rxn} for the reaction:

$$CH_{4(g)} + 2\ O_{2(g)} \rightleftharpoons CO_{2(g)} + 2\ H_2O_{(g)}$$

The Mole Says

Just like enthalpy, the free energy of formation for a pure element is zero.

Given the following information:

$\Delta G^\circ_f(CH_{4(g)}) = -50.8$ kJ/mol

$\Delta G^\circ_f(O_{2(g)}) = 0$ kJ/mol

$\Delta G^\circ_f(CO_{2(g)}) = -394.4$ kJ/mol

$\Delta G^\circ_f(H_2O_{(g)}) = -228.6$ kJ/mol

Solution: To solve this problem, we need to subtract the free energies of the reactants from those of the products.

Reactants:

ΔG_f^o for 1 mol CH_4 = 1 mol × –50.8 kJ/mol = –50.8 kJ

ΔG_f^o for 2 mol O_2 = 2 mol × 0 kJ/mol = 0 kJ

Total: –50.8 kJ

Products:

ΔG_f^o for 1 mol CO_2 = 1 mol × –394.4 kJ/mol = –394.4 kJ

ΔG_f^o for 2 mol H_2O = 2 mol × –228.6 kJ/mol = –457.2 kJ

Total: –851.6 kJ

$\Delta G^o_{rxn} = \Delta G^o_{products} - \Delta G^o_{reactants}$

= –851.6 kJ – (–50.8 kJ)

= –800.8 kJ

The Dependence of Free Energy on Temperature

As we saw before, the entropy contribution to ΔG in the equation for free energy changes when the temperature changes (the –TΔS term in ΔG = ΔH–TΔS). As a result, the standard free energy of a process won't necessarily be the same thing as the free energy at different temperatures.

We can qualitatively predict whether a reaction will be spontaneous by using the following table, which shows how ΔH and ΔS affect ΔG, particularly as the temperature changes.

ΔH	ΔS	ΔG	**Is It Spontaneous?**
–	+	always –	Yes, at all temperatures
+	–	always +	No, at all temperatures
+	+	+ at low T	No, at low temperatures
		– at high T	Maybe, at high temperatures
–	–	– at low T	Maybe, at low temperatures
		+ at high T	No, at high temperatures

Quantitatively, we can find the change in free energy for the process by finding the standard entropy and enthalpy of the process, then use the equation $\Delta G = \Delta H - T\Delta S$ to incorporate the change in temperature.

Example: Find the change in free energy for the following reaction at 773 K:

$$2\ H_{2(g)} + O_{2(g)} \rightleftharpoons 2\ H_2O_{(g)}$$

Given the following information:

$$\Delta H^\circ_{rxn} = -483.6\ kJ$$

$$\Delta S^\circ_{rxn} = -88.8\ J/K$$

Solution: Before we can solve this problem, we need to convert ΔS°_{rxn} to kJ/K so the units of both entropy and enthalpy are in kJ. To do this, we divide ΔS°_{rxn} by 1,000 to get $\Delta S^\circ_{rxn} = -0.0888$ kJ/K.

Placing the values for entropy, enthalpy, and temperature into the equation for free energy, we find that:

$$\Delta G^\circ = \Delta H^\circ - T\Delta S^\circ$$

$$= -483.6\ kJ - (773K \times -0.0888\ kJ/K)$$

$$= -483.6\ kJ + 68.6\ kJ$$

$$= -415.0\ kJ$$

You've Got Problems

Problem 2: Determine the free energy change for the following reaction at 500 K:

$$C_{(s)} + 2\ H_{2(g)} \rightleftharpoons CH_{4(g)}$$

Given the following information:

$$\Delta H^\circ_{rxn} = -74.8\ kJ$$

$$\Delta S^\circ_{rxn} = -80.6\ J/K$$

Relating Free Energy to the Equilibrium Constant

By looking at the magnitude of ΔG° for a process, we can get a pretty good feel for what the equilibrium constant, K (Chapter 22), is for the reaction.

- When $\Delta G^\circ = 0$, the reaction is at equilibrium. At equilibrium, the equilibrium constant K is equal to 1. As a result, when $\Delta G^\circ = 0$, $K = 1$.

- When ΔG° is negative, the reaction is proceeding spontaneously. When a reaction proceeds spontaneously in the forward direction, the equilibrium constant for the process is greater than one. As a result, when $\Delta G^\circ < 0$, $K > 1$.
- When ΔG° is positive, the reaction doesn't proceed spontaneously in the forward direction, but does spontaneously move from products to reactants. When a reaction goes backward, the equilibrium constant K is less than one. As a result, when $\Delta G^\circ > 0$, $K < 1$.

What does this mean? If we see that the equilibrium constant for a reaction is very high (for example, 3,387), the free energy for the reaction will be very negative. This, in turn, tells us that the reaction will proceed spontaneously. Likewise, if the equilibrium constant for a reaction is very low (for example, 0.00000034), the free energy for the reaction will be very positive, indicating a nonspontaneous process.

The Least You Need to Know

- Spontaneous processes can occur without any outside intervention.
- Increases in entropy (randomness) are a driving force for spontaneous processes.
- The free energy of a process is the final word on whether it can be spontaneous, and takes into account entropy, enthalpy, and temperature.
- The free energy and equilibrium constant of a given process are closely related to one another.

Appendix

Solutions to "You've Got Problems"

Chapter 2

1. (a) 0.0000075 meters is equal to 7.5×0.000001 meters, or 7.5 µm.

   (b) 25,000,000 grams is equal to $25 \times 1,000,000$ grams, or 25 Mg.

2. $160 \text{ pounds} \times \frac{1 \text{ kilogram}}{2.21 \text{ pounds}} = 72 \text{ kilograms}$

3. This problem needs to be solved in two steps. The first step involves converting 25 miles to kilometers using the conversion factor provided in the problem, and the second step involves converting kilometers to meters using the prefixes discussed earlier in the chapter.

   $25 \text{ miles} \times \frac{1.6 \text{ kilometers}}{1 \text{ miles}} = 40. \text{ kilometers}$

   $40. \text{ kilometers} \times \frac{1000 \text{ meters}}{1 \text{ kilometers}} = 4.0 \times 10^4 \text{ meters}$

4. (a) 2.490 grams has four significant figures.

   (b) 1010 grams has three significant figures.

   (c) 0.01010 grams has four significant figures.

5. (a) 5.6 kilometers

   (b) 7.6 grams

Chapter 4

1. (a) 6 protons, 8 neutrons, 6 electrons

   (b) 15 protons, 16 neutrons, 15 electrons

   (c) 7 protons, 7 electrons, cannot determine the number of neutrons because the isotope isn't specified.

2. Average atomic mass = (10.013 amu)(0.199) + (11.009)(0.801)

   = 1.99 amu + 8.82 amu

   = 10.81 amu

Chapter 5

1. Because the value of *l* determines the type of orbital, these quantum numbers describe a p-orbital.

2. (a) Ga: $1s^2 2s^2 2p^6 3s^2 3p^6 4s^2 3d^{10} 4p^1$

   (b) Ir: $1s^2 2s^2 2p^6 3s^2 3p^6 4s^2 3d^{10} 4p^6 5s^2 4d^{10} 5p^6 6s^2 4f^{14} 5d^7$

3. (a) Y: $[Kr]5s^2 4d^1$

   (b) Po: $[Xe]6s^2 4f^{14} 5d^{10} 6p^4$

4. 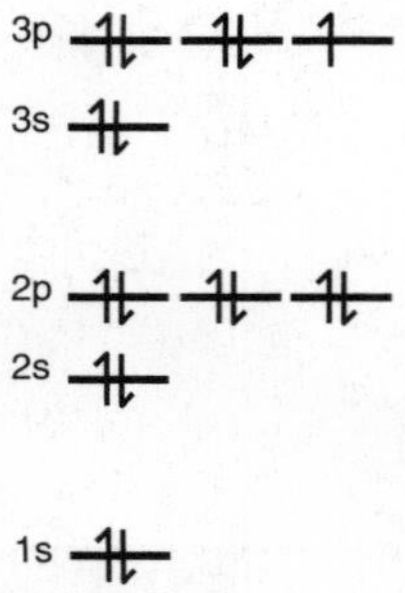 3p ↑↓ ↑↓ ↑

   3s ↑↓

   2p ↑↓ ↑↓ ↑↓

   2s ↑↓

   1s ↑↓

Chapter 6

1. (a) Turkey stuffing is a heterogeneous mixture.

   (b) Sugar water is a homogeneous mixture.

   (c) Chunky peanut butter is a heterogeneous mixture.

Chapter 7

1. electronegativity: Rb < Sn < P < O < F

   atomic radius: F < O < P < Sn < Rb

Chapter 8

1. (a) Magnesium (Mg) will have a charge of +2.

   (b) Aluminum (Al) will have a charge of +3.

   (c) Bromine (Br) will have a charge of –1.

2. (a) ionic

   (b) not ionic

   (c) not ionic

3. (a) Na_2SO_4 is sodium sulfate (remember, not everything needs a Roman numeral!)

   (b) Cu_2O is copper(I) oxide.

   (c) $CoCO_3$ is cobalt(II) carbonate.

   (d) NH_4Cl is ammonium chloride.

4. (a) Lithium acetate is $LiC_2H_3O_2$.

   (b) Sodium nitrate is $NaNO_3$.

   (c) Chromium(VI) sulfate is $Cr(SO_4)_3$.

   (d) Zinc phosphate is $Zn_3(PO_4)_2$.

Chapter 9

1. $H\cdot \quad \cdot H \quad \cdot\ddot{N}\cdot \quad H\cdot \longrightarrow H{:}\ddot{N}{:}H$ with H below N

2. (a) PCl_3 is phosphorus trichloride.

   (b) CO is carbon monoxide.

   (c) SF_6 is sulfur hexafluoride.

3. (a) HBr

   (b) SiO_2

   (c) OCl_2

Chapter 10

1. (a) $H-\ddot{N}-H$ with H bonded below N

   (b) $\ddot{O}=Si=\ddot{O}$

   (c) $:\ddot{O}-H$ with charge -1

2. $\left[H-C(=\ddot{O}:)-\ddot{O}:^{-1} \longleftrightarrow H-C(-\ddot{O}:^{-1})=\ddot{O}: \right]$

3. (a) SiO_2 is linear, has sp hybridization, and a bond angle of 180°.

   (b) CH_2O is trigonal planar, has sp^2 hybridization, and a bond angle of 120°.

   (c) OF_2 is bent, has sp^3 hybridization, and a bond angle of 104.5°.

Chapter 11

1. (a) Na_2SO_4 has a molar mass of 142.05 g/mol.

   (b) Nitrogen trichloride (NCl_3) has a molar mass of 120.36 g/mol.

   (c) Fluorine (F_2) has a molar mass of 38.00 g/mol.

2. Solving this problem requires two calculations. The first involves converting molecules of $POCl_3$ to moles, and the second requires converting moles of $POCl_3$ to grams. Each step is shown next.

3. $$4.3 \times 10^{22} \cancel{\text{molecules } POCl_3} \times \frac{1 \text{ mole } POCl_3}{6.02 \times 10^{23} \cancel{\text{molecules } POCl_3}} = 0.071 \text{ moles } POCl_3$$

   $$0.071 \cancel{\text{moles } POCl_3} \times \frac{153.32 \text{ grams } POCl_3}{1 \cancel{\text{mole } POCl_3}} = 11 \text{ grams } POCl_3$$

3. There is one silver atom in one mole of $AgNO_3$ with a mass of 107.87 grams. The molar mass of $AgNO_3$ is 169.88 grams. As a result, the mass percent of silver in $AgNO_3$ is equal to (107.87 grams/169.88 grams) × 100%, or 63.498%. When we find 63.498% of 25 grams of silver nitrate, our answer is 16 grams of silver.

Chapter 13

1. δ– δ+ δ–

   :C̈l: – P̈ – :C̈l:

   |

   :Cl:

   δ–

2. (a) dipole-dipole forces

   (b) London dispersion forces

   (c) hydrogen bonding

3. $CF_4 < PF_3 < HF$

Chapter 14

1. (a) yes

   (b) yes

   (c) no

2. Because the molar mass of acetic acid is 60.0 g/mol, we have 120/60.06 = 2.0 moles of acetic acid. Since we have three liters of solution, the molarity is equal to 2.0 mol/3.0 L = 0.67 M.

3. 45.0 grams $Ca(C_2H_3O_2)_2$ is equal to 0.284 moles

   560 mL H_2O = 0.56 kg H_2O

   m = 0.284 moles / 0.56 kg H_2O = 0.51 m

4. $$\chi_{water} = \frac{15.0 \text{ moles water}}{15.0 \text{ moles water} + 4.5 \text{ moles isopropanol}} = 0.77$$

5. M_1 = 0.500 M, M_2 = 0.125 M, V_1 is unknown, V_2 = 750 mL. Inserting these values into the equation $M_1V_1 = M_2V_2$, we get:

   $(0.500 \text{ M})V_1 = (0.125 \text{ M})(750 \text{ mL})$

   V_1 = 190 mL

   I will need 190 mL of 0.500 M NaCl to make the desired solution.

Chapter 15

1. This is a trick question! There's no such thing as an ideal gas! We pretend as if gases are ideal so that we may predict how they behave, but in reality, no gases truly qualify as ideal.

2. $$u_{rms} = \sqrt{\frac{3(8.314J/molK)(273K)}{0.00101g/mol}} = 2.60x10^3 m/\sec$$

3. $$\frac{r_{He}}{r_{H_2}} = \sqrt{\frac{0.00202kg}{0.00400kg}} = 0.710$$

   This indicates that the helium will effuse from the balloon at a rate 0.710 times as quickly as the hydrogen. If hydrogen effuses in 16.0 hours, helium will effuse in 16.0 hrs/0.710 = 22.5 hours.

Chapter 16

1. P_1 = 1.00 atm, V_1 = 1500 L, P_2 = 450 atm, V_2 = × L. Plugging these numbers into Boyle's Law, we find that the volume of the gas is:

   (1.00 atm)(1500 L) = (450 atm)(x L)

   x = 3.3 L

2. V_1 = 2.5 L, T_1 = 298 K, V_2 = X L, T_2 = 323 K. Using Charles's Law, we find that the new volume of the gas is:

   $$\frac{2.5L}{298K} = \frac{xL}{323K}$$

   $x = 2.7L$

3. P_1 = 20.0 atm, T_1 = 293 K, P_2 = 35.0 atm, T_2 = × K. Using Gay-Lussac's Law, the maximum temperature we can heat a propane tank to is:

   $$\frac{20.0atm}{293K} = \frac{35.0atm}{xK}$$

   x = 513 K, or 240° C

4. P_1 = 1.00 atm, V_1 = 5.00 × 10^3 L, T_1 = 293 K, P_2 is unknown, V_2 = 1,170 L, T_2 is 275 K. Using the combined gas law:

   $$\frac{(1.00atm)(50{,}000L)}{(293K)} = \frac{(P_2)(1{,}170L)}{(275)}$$

   P_2 = 40.1 atm

5. P = 1.0 atm, V = 1,100 L, n is unknown, R = 0.08206 L atm/mol K, T = 523 K. Using the ideal gas law:

   (1.0 atm)(1,100 L) = n (0.08206 L atm/mol K)(523 K)

   n = 26 moles

6. (a) To find the number of moles of each gas, we use the ideal gas law using the values for each individually. After all, we assume that the gases don't interact with each other at all.

   To find moles of oxygen, use P = 5.0 atm, V = 55 L, n = x moles, R = 0.08206 L.atm/mol.K, T = 288 K in the ideal gas law:

(5.0 atm)(55 L) = (x moles O2)(0.08206 L.atm/mol.K)(288 K)

x = 12 moles O_2

Likewise, to find moles of nitrogen, we use P = 8.0 atm, V = 55 L, n = x moles, R = 0.08206 L·atm/mol·K, T = 288 K in the ideal gas law:

(8.0 atm) (55 L) = (x moles O_2)(0.08206 L·atm/mol·K)(288 K)

x = 19 moles N_2

b) To find the total pressure, remember Dalton's Law says that the total pressure in the container is equal to the sum of the partial pressures of each gas. Because the partial pressure of oxygen is 5.0 atm and the partial pressure of nitrogen is 8.0 atm, the total pressure in the container is (5.0 atm + 8.0 atm =) 13.0 atm.

Chapter 17

1. From the fact that it evaporates more quickly you can determine that rubbing alcohol has a much higher vapor pressure than iced tea.

2. The molality of the $ZnCl_2$ solution is found by dividing the moles of solute by kilograms of solvent, or 2.5 mol/2.0 kg = 1.3 m. However, since $ZnCl_2$ is an ionic compound that forms three ions when dissolved, the effective molality of the solution is three times that, or 3.9 m.

   Placing this into the equation for boiling point elevation:

   $\Delta T = K_b m_{solute}$

   = (0.051 °C/m)(3.9 m)

   0.20° C

   Because the boiling point is raised by 0.20° C over the original boiling point of 100° C, the boiling point of the solution is 100.20° C.

3. Using the equation $\Delta T_f = K_f\, m_{solute}$ and the value 1.50° C for ΔT_f and the value 1.86°C/m for K_f, we find that:

   1.50° C = (1.86°C/m)(x m)

   x = 0.806 m

   However, this value is the molality of the number of particles in this solution. Since we are making this solution using NaOH, which breaks into two particles (the Na^+ and OH^- ions), the effect of this solute on the melting point is twice as strong as you might expect. Dividing the value 0.806 m by two, we find the actual concentration of NaOH to be 0.403 m.

4. Using Figure 17.8, the phase diagram of water, you find that the conditions of –10° C and 0.9 atm put you in the "solid" range. To move directly into the "gas" area of this graph (remember, sublimation is when a material turns from a solid directly into a gas) you'll need to move lower on the graph by lowering the pressure of the ice.

Chapter 18

1. (a) $1\ CaCl_2 + 2\ AgNO_3 \rightleftharpoons 2\ AgCl + 1\ Ca(NO_3)_2$

   (b) $3\ (NH_4)_2CO_3 + 2\ FeBr_3 \rightleftharpoons 1\ Fe_2(CO_3)_3 + 6\ NH_4Br$

   (c) $1\ P_4 + 5\ O_2 \rightleftharpoons 2\ P_2O_5$

2. (a) $Pb(NO_3)_{2(aq)} + 2\ KI_{(aq)} \rightleftharpoons 1\ PbI_{2(s)} + 2\ KNO_{3(aq)}$

   (b) $4\ Fe_{(s)} + 3\ O_{2(g)} \overset{\Delta}{\rightleftharpoons} 2\ Fe_2O_{3(s)}$

3. (a) Double displacement reaction

   (b) Single displacement reaction

   (c) Acid-base reaction

   (d) Combustion reaction

4. (a) $2\ NaOH + 1\ H_2SO_4 \rightleftharpoons 1\ Na_2SO_4 + 2\ H_2O$

   (b) $2\ NH_3 + 3\ I_2 \rightleftharpoons 2\ NI_3 + 3\ H_2$

   (c) $2\ C_3H_8O + 9\ O_2 \rightleftharpoons 6\ CO_2 + 8\ H_2O$

Chapter 19

1. $125\ \cancel{g\ CaCl_2} \times \frac{1\ \cancel{mol\ CaCl_2}}{111.1\ \cancel{g\ CaCl_2}} \times \frac{1\ \cancel{mol\ Ca(OH)_2}}{1\ \cancel{mol\ CaCl_2}} = 1.12\ mol\ Ca(OH)_2$

2. This problem requires two calculations—one for determining the mass of lead iodide that can be formed from 115 grams of lead nitrate and one for determining the mass of lead iodide that can be formed from 265 grams of potassium

iodide. The smaller of the two answers will be the answer for this question, corresponding to the limiting reactant.

$$115 \text{ g Pb(NO}_3)_2 \times \frac{1 \text{ mol Pb(NO}_3)_2}{331.22 \text{ g Pb(NO}_3)_2} \times \frac{1 \text{ mol PbI}_2}{1 \text{ mol Pb(NO}_3)_2} \times \frac{461.00 \text{ g PbI}_2}{1 \text{ mol PbI}_2} = 160. \text{ g PbI}_2$$

$$265 \text{ g KI} \times \frac{1 \text{ mol KI}}{166.00 \text{ g KI}} \times \frac{1 \text{ mol PbI}_2}{2 \text{ mol KI}} \times \frac{461.00 \text{ g PbI}_2}{1 \text{ mol PbI}_2} = 368 \text{ g PbI}_2$$

From these calculations, we can see that lead nitrate is the limiting reactant and that 160. g PbI_2 will be formed in this reaction.

3. The amount of excess reactant left over is:

$= 265 \text{ g KI} - (265 \text{ g KI})(160. \text{ g PbI}_2/368 \text{ g PbI}_2)$

$= 265 \text{ g KI} - 115 \text{ g KI}$

$= 150. \text{ g KI}$

Chapter 20

1.

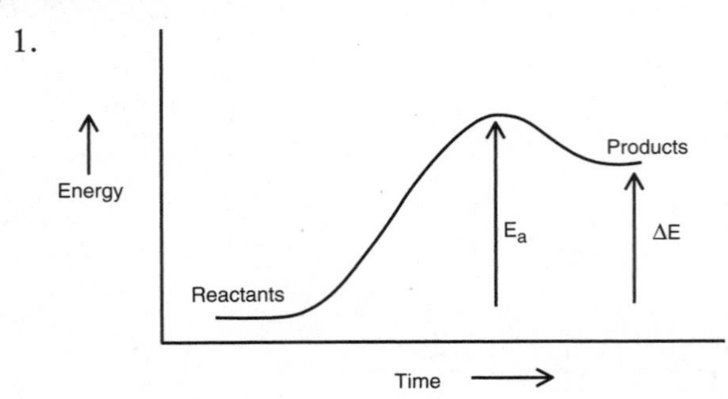

2. Chilling egg salad sandwiches causes there to be less energy present for the spoiling "reaction" to occur and make them rancid.
3. When you grind coffee beans you break the beans into smaller particles, giving them a larger surface area over which water can be converted into coffee.

Chapter 21

1. From the data provided, you can see that the reaction rate is four times faster when the concentration of A is doubled ($2.00 \times 10^{-5} / 5.00 \times 10^{-6} = 4$) and that the rate stays the same when the concentration of B is doubled. As a result, the reaction is second order in [A] and zeroth order in [B], making the rate law:

   Rate = $k[A]^2$

   To find the rate constant, let's place use the data from experiment 1 into this equation (it doesn't matter which experiment you use—they'll all give the same answer):

   5.00×10^{-6} M/s = $k(0.0100\ M)^2$

   $k = 5.00 \times 10^{-2}$/M·s

   Finally, we can determine the rate of the reaction when [A] = 0.0150 M and [B] = 0.250 M by placing this information into the rate equation with the value for the rate constant that we determined previously.

   Rate = $(5.00 \times 10^{-2}/M{\cdot}s)(0.0150\ M)^2$

   $= 1.13 \times 10^{-5}$ M/s

2. You can verify that this is a first order reaction in [A] by graphing ln[A] vs. time—this graph produces a straight line:

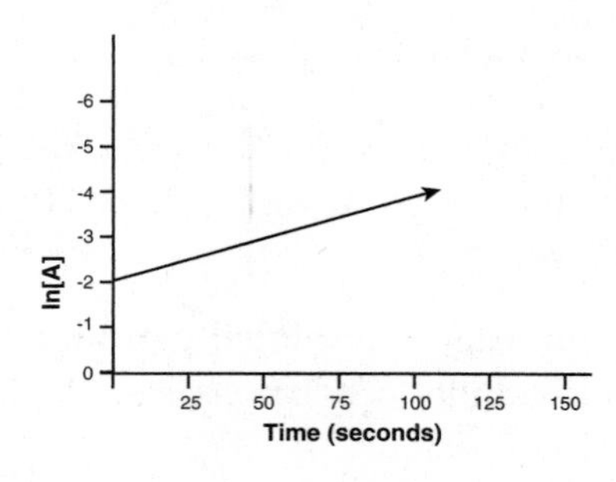

   The slope of this line is –k, or –(–0.0139/M·s) or 0.139/M·s.

3. Solving the equation relating k and half-life using the information in this problem, we get:

$$t_{1/2} = \frac{0.693}{k}$$

$$65\,\text{sec} = \frac{0.693}{k}$$

$$k = 0.011\,/\,\text{sec}$$

4. $$\ln\left(\frac{k_2}{k_1}\right)=\left(\frac{E_a}{R}\right)\left(\frac{1}{T_1}-\frac{1}{T_2}\right)$$

$$\ln\left(\frac{122L/mol\cdot s}{3.40L/mol\cdot s}\right)=\left(\frac{E_a}{8.31J/K\cdot mol}\right)\left(\frac{1}{1073K}-\frac{1}{1123K}\right)$$

$$3.58=\left(\frac{E_a}{8.31J/K\cdot mol}\right)(4.15x10^{-5}/K)$$

$$E_a=717{,}000J/mol$$

Chapter 22

1. (a) The equilibrium expression is:

$$K_c=\frac{[HI]^2}{[H_2][I_2]}$$

(b) Making the chart that shows how the concentrations of each species change, we get.

Species	Initial Concentration	Change	Final Concentration
H_2	2.0 M	–x	2.0 – x
I_2	2.0 M	–x	2.0 – x
HI	0 M	2x	2x

The reason that the change in HI is 2x instead of x as in our earlier example is that by the equation, for every one molecule of H_2 or I_2 that react, we get two molecules of HI formed.

Substituting these values and the solubility product constant we were given into the equation for the solubility product constant yields:

$$5=\frac{[2x]^2}{[2.0-x][2.0-x]}$$

which, combining the terms in the denominator together, is the same as:

$$5=\frac{[2x]^2}{[2.0-x]^2}$$

If we take the square root of both sides, we get:

$$2.24=\frac{2x}{2.0-x}$$

And solving for x, we get an answer of 1.06 M. Note that in this example, we don't need to use the approximation that x is negligible when compared to 2 M to obtain an exact answer.

2. $PbBr_2$ dissociates via the following process:

   $PbBr_{2(s)} \rightleftharpoons Pb^{+2}_{(aq)} + 2\ Br^-_{(aq)}$

   making our K_{sp} expression:

   $K_{sp} = [Pb^{+2}][Br^-]^2$

   When lead bromide dissociates, we find that the concentration of the bromide ion will be twice that of the lead ion because two bromide ions are created whenever $PbBr_2$ breaks apart. As a result, we can say that the concentration of Pb^{+2} in a saturated solution is x, and the concentration of Br^- in a saturated solution is 2x. Placing these in our expression, we find that:

   $K_{sp} = 2.1 \times 10^{-6} = [x][2x]^2$

   $2.1 \times 10^{-6} = 4x^3$

   $x = 8.1 \times 10^{-3}$ M

   As a result, the concentration of Pb^{+2} will be 8.1×10^{-3} M, and the concentration of Br^- will be twice that, or 1.6×10^{-2} M.

3. (a) The reaction will shift toward products.

   (b) The reaction will shift toward reactants.

   (c) The reaction will shift toward products.

   (d) The reaction will shift toward products.

Chapter 23

1. H_2SO_4 is an acid and HSO_4^{-1} is its conjugate base.

   HPO_4^{2-} is a base and $H_2PO_4^{-1}$ is its conjugate acid.

2. The molarity of this solution is found by dividing the number of moles of HCl by the number of liters of water. Because 1.20 grams of HCl is 0.0329 moles and the volume is 0.450 L, the molarity of this solution is 0.0731 M.

   To find the pH, simply put this number into the equation $pH = -\log[H^+]$ to get an answer of 1.14.

3. Problem 2: What's the pH of a 0.750 M formic acid (HCO_2H) solution? K_a (HCO_2H) = 1.77×10^{-4} M.

   The equilibrium formed when formic acid dissociates in water is expressed by the following equation:

   $HCO_2H \rightleftharpoons HCO_2^- + H^+$

   $$K_a = \frac{[HCO_2^-][H^+]}{[HCO_2H]}$$

   Initially, the concentration of formic acid is 0.750 M. However, at equilibrium, some of the formic acid will have dissociated, so the concentration of the formic acid will have decreased to "0.750 – x" M.

   For every molecule of formic acid that dissociates, one formate ion and one hydronium ion will be formed. As a result, the quantity of each ion formed will be equal to "x." Replacing the quantities in the expression, we get:

   $$1.77x10^{-4} = \frac{[x][x]}{[0.750 - x]}$$

   Because formic acid is a weak acid, we'll assume that "x" is a small value compared to 0.750 M. We can simplify the previous equation as:

   $$1.77x10^{-4} = \frac{[x][x]}{[0.750]}$$

   x = 0.0115 M

   To find the pH, we'll place this value for $[H^+]$ into the equation for pH to get:

   $pH = -\log[H^+]$

   $pH = -\log(0.0115)$

   pH = 1.94

4. $K_w = [H^+][OH^-]$

   $1.00 \times 10^{-14} = [H^+][0.00340\ M]$

   $[H^+] = 2.94 \times 10^{-12}$ M

   $pH = -\log[H^+]$

   $pH = -\log[2.94 \times 10^{-12}]$

   pH = 11.5

5. Using $M_1V_1 = M_2V_2$, we find:

   $(0.500\ M)(245\ mL) = M_2\ (175\ mL)$

   $M_2 = 0.700\ M$

   The concentration of the NaOH solution is 0.700 M.

6. Inserting the values given in the problem into the Henderson-Hasselbalch equation we get:

   $$pH = -\log K_a + \log \frac{[base]}{[acid]}$$

   $$pH = -\log(1.8 \times 10^{-4}) + \log \frac{[0.75M]}{[0.50M]}$$

   $$pH = 3.7 + 0.18$$

   $$pH = 3.9$$

Chapter 24

1. (a) Because bromine has a higher electronegativity than phosphorus, the oxidation state of bromine is the same as if it were anionic, or –1 each. Because three bromine atoms each have a –1 oxidation state, phosphorus must have a +3 oxidation state for the sum of the oxidation states to be zero.

   (b) NaOH is an ionic compound. As a result, Na has a charge of +1, and the sum of the charges on the hydroxide ion is –1. As hydrogen is bonded to a nonmetal, the charge on hydrogen is +1. For the overall charge on the hydroxide ion to be –1, oxygen must have an oxidation state of –2.

   (c) Each hydrogen has a +1 oxidation state. Because oxygen is more electronegative than sulfur, each oxygen must have an oxidation state of –2. To make the sum of the oxidation states of all atoms in this molecule zero, sulfur must have an oxidation state of +6 $[+6 + (2 \times +1) + (4 \times -2)] = 0$).

2. On the left side of the equation, carbon has a –4 oxidation state, hydrogen has a +1 oxidation state, and oxygen has an oxidation state of 0. On the right side of the equation, carbon has an oxidation state of +4, all of the oxygen atoms have an oxidation state of –2, and hydrogen has an oxidation state of +1. As a result, methane is the reducing agent (because carbon is oxidized) and oxygen is the oxidizing agent (because oxygen is reduced).

3. Breaking this up into the half-reactions, we get:

   $As_2O_3 \rightarrow H_3AsO_4$

   $NO_3^- \rightarrow NO$

   When we balance the species that are oxidized or reduced, we get:

   $As_2O_3 \rightarrow 2\ H_3AsO_4$

   $NO_3^- \rightarrow NO$ (No change)

   Adding water to balance the oxygen atoms, we get:

   $As_2O_3 + 5\ H_2O \rightarrow 2\ H_3AsO_4$

   $NO_3^- \rightarrow NO + 2\ H_2O$

   Adding H^+ to balance hydrogen:

   $As_2O_3 + 5\ H_2O \rightarrow 2\ H_3AsO_4 + 4\ H^+$

   $NO_3^- + 4\ H^+ \rightarrow NO + 2\ H_2O$

   When we add electrons to neutralize both sides of each equation, we end up with:

   $As_2O_3 + 5\ H_2O \rightarrow 2\ H_3AsO_4 + 4\ H^+ + 4\ e^-$

   $NO_3^- + 4\ H^+ + 3\ e^- \rightarrow NO + 2\ H_2O$

   To balance the number of electrons transferred in both equations, we need to multiply the coefficients in the first half-reaction by 3 and the coefficients in the second half-reaction by 4:

   $3\ As_2O_3 + 15\ H_2O \rightarrow 6\ H_3AsO_4 + 12\ H^+ + 12\ e^-$

   $4\ NO_3^- + 16\ H^+ + 12\ e^- \rightarrow 4\ NO + 8\ H_2O$

   Add them up and get one big reaction:

   $3\ As_2O_3 + 15\ H_2O + 4\ NO_3^- + 16\ H^+ + 12\ e^- \rightarrow 6\ H_3AsO_4 + 12\ H^+ + 12\ e^- + 4\ NO + 8\ H_2O$

   When we cancel the terms that appear on both sides, we get our final answer of:

   $3\ As_2O_3 + 7\ H_2O + 4\ NO_3^- + 4\ H^+ \rightleftharpoons 6\ H_3AsO_4 + 4\ NO$

4. The oxidation reaction is $Al \rightarrow Al^{+3} + 3\ e^-$. Because the standard reduction potential for this process is –1.66 V, the half-cell potential for the oxidation is +1.66 V. For the reduction reaction $Fe^{+2} + 2\ e^- \rightarrow Fe$, the standard reduction potential is –0.44 V. As a result, the standard cell potential for this voltaic cell is 1.66 V – 0.44 V = 1.22 V.

Chapter 25

1. 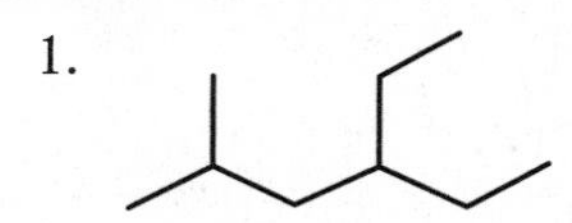

2.

3.

Chapter 26

1. a) $^{108}_{48}Ag \rightarrow ^{0}_{-1}e + ^{108}_{46}Pd$

   b) $^{216}_{86}Rn \rightarrow ^{4}_{2}He + ^{212}_{84}Po$

2. $t_{1/2} = \dfrac{0.693}{k}$

   $k = \dfrac{0.693}{t_{1/2}} = \dfrac{0.693}{75 \text{ years}} = 0.0092 \text{ / year}$

Chapter 27

1. The equation for this reaction is:

   $C_{12}H_{22}O_{11} + 12\ O_2 \leftrightarrow 12\ CO_{2(g)} + 11\ H_2O_{(l)}$

   Products:

   ΔH°_{f} for 12 mol CO_2 = 12 mol × –393.5 kJ/mol = –4722 kJ

   ΔH°_{f} for 11 mol H_2O = 11 mol × –285.8 kJ/mol = –3144 kJ

   Total: = –7866 kJ

Reactants:

ΔH°_f for 1 mol $C_{12}H_{22}O_{11(s)}$ = 1 mol × –2221 kJ/mol = –2221 kJ

ΔH°_f for 12 mol $O_{2(g)}$ = 12 mol × 0.00 kJ/mol = 0.00 kJ

Total: = –2221 kJ

$\Delta H^{\circ}_{rxn} = \Delta H^{\circ}_f$ (products)—ΔH°_f (reactants)

ΔH°_{rxn} = –7866 kJ – (–2221 kJ)

ΔH°_{rxn} = –5645 kJ

2. The way to solve this problem is to take the first equation and reverse it to get:

   $CO_{2(g)} \rightleftharpoons C(\text{diamond}) + O_{2(g)}$ ΔH_{rxn} = +395.4 kJ

   Notice that the sign on ΔH_{rxn} is reversed.

   Upon adding this to the second equation, we get the following:

   $\cancel{CO_{2(g)}} + C(\text{graphite}) + \cancel{O_{2(g)}} \rightleftharpoons C(\text{diamond}) + \cancel{O_{2(g)}} + \cancel{CO_{2(g)}}$

   C (graphite) $\rightleftharpoons$ C (diamond)

   ΔH_{rxn} for this process is +395.4 kJ – 393.5 kJ = + 1.9 kJ.

3. The heat of combustion of one mole of naphthalene is 5,154 kJ. However, I only have 1.00 g naphthalene, or 0.00780 mol, so the amount of energy I would expect to be released is 5,154 kJ/mol × 0.00780 mol = 40.2 kJ.

   Using the equation $\Delta H = mC_p\Delta T$ and solving for ΔT, I find that:

   40,200 J = (1,500 g)(4.184 J/g°C)(ΔT)

   ΔT = 6.41° C

Chapter 28

1. To solve this problem, we need to subtract the entropies of the reactants from those of the products.

   Reactants:

   ΔS° of 1 mol $FeCl_3$ = 1 mol × 142.3 J/mol K = 142.3 J/K

   ΔS° of 3 mol Na = 3 mol × 51.3 J/mol K = 153.9 J/K

   Total: 296.2 J/K

Products:

ΔS° of 1 mol Fe = 1 mol × 27.2 J/mol K = 27.2 J/K

ΔS° of 3 mol NaCl = 3 mol × 72.3 J/mol K = 216.9 J/K

Total: 244.1 J/K

$\Delta S^\circ_{rxn} = \Delta S^\circ_{products} - \Delta S^\circ_{reactants}$

= 244.1 J/K – 296.2 J/K

= –52.1 J/K

2. As in the example in the chapter, we need to convert the units of ΔS°_{rxn} to kJ/K so the units of entropy and enthalpy are the same. Dividing by 1,000, ΔS°_{rxn} = –0.0806 kJ/K.

   Placing the values for entropy, enthalpy, and temperature into the equation for free energy:

   $\Delta G = \Delta H - T\Delta S$

   = –74.8 kJ – (500 K × –0.0806 kJ/K)

   = –74.8 kJ + 40.3 kJ

   = –34.5 kJ

Appendix B

Glossary

absorption (1) In spectroscopy, when light is used to push an electron from a ground state to an excited state. (2) When a chemical is soaked up by a material such as water being absorbed into paper towels.

accuracy When a measured value is close to the actual value. Accurate measurements are also precise.

acid Any material that can accept a pair of electrons. In aqueous solutions, those with pH < 7.00.

acid-base reaction An electron pair is donated by a base to an acceptor acid. In an aqueous solution, the reaction of H^+ and OH^- yields water.

acid dissociation constant (K_a) The constant that describes the equilibrium position for the dissolution of an acid in water.

activation energy The minimum amount of energy that's required for the reactants of a chemical reaction to form products.

adsorption When a chemical is stuck to the surface of a material.

alkali metals All elements in group 1 of the periodic table, except hydrogen. Alkali metals are the most reactive group of metals.

alkaline earth metals Reactive elements in group 2 of the periodic table.

alkane A hydrocarbon that contains only single carbon-carbon bonds. Also called a "saturated hydrocarbon."

alkene A hydrocarbon that contains at least one double bond.

alkyne A hydrocarbon that contains at least one triple bond.

alloy A metal in which several elements are present.

alpha decay When a nucleus breaks apart, emitting a helium nucleus, which is called an "alpha particle" in this context.

amorphous solid A solid material in which the molecules have no long-range order.

amu Atomic mass unit, equivalent to $\sim 1.67 \times 10^{-27}$ kg.

angular momentum quantum number Denoted by "l," it determines the shape and type of the orbital. Possible values are 0, 1, 2, … (n-1).

anhydride A compound that forms an acid or base when combined with water.

anion A negatively charged atom or group of atoms.

anode The electrode where oxidation occurs.

aromatic hydrocarbon A hydrocarbon containing alternating single C-C and double C-C bonds in a ring. The best known aromatic hydrocarbon is benzene (C_6H_6).

Arrhenius acid A compound that forms hydronium (H^+) ions in water.

Arrhenius base A compound that forms the hydroxide ion in water.

atmosphere (atm) A unit of pressure equal to the average atmospheric pressure at sea level.

atom The smallest chunk of an element with the same properties as larger chunks of that element.

atomic mass The sum of the number of protons and number of neutrons in the nucleus of an atom, denoted by the symbol A.

atomic number The number of protons in an element, denoted by the symbol Z.

atomic radius One half the distance between the nuclei of two bonded atoms of the same element.

atomic symbol The symbol for each element found on the periodic table.

average atomic mass The weighted average of the masses of all of the isotopes of an element.

Avogadro's Law The molar volumes of all ideal gases are the same.

Avogadro's number 6.02×10^{23}.

base Any molecule that can donate a pair of electrons to form a bond. In aqueous solutions, those with $pH > 7.00$.

beta decay When an electron (called a "beta particle" in this context) is emitted during the radioactive decay of an atomic nucleus.

binding energy The energy due to the mass defect of an atom. It's responsible for holding the nucleus together.

Brønsted-Lowry acid A compound that gives H^+ ions to other compounds.

Brønsted-Lowry base A compound that accepts H^+ ions from other compounds.

buffer A solution consisting of a weak acid and its conjugate base that resists changes in pH when acid or base is added to it.

buffering capacity The quantity of acid or base that can be added to a buffered solution before the pH undergoes significant change.

calorimetry The process by which the energy change of a process is experimentally determined.

catalyst A material that increases the rate of a chemical reaction without being consumed.

cathode The electrode at which reduction occurs.

cation An atom or group of atoms with positive charge.

cell potential A measure of the electromotive force that drives electrons in a voltaic cell.

chromatography A method of separating a mixture in which the components are passed through a third material. The affinity of each component of the mixture to stick to this third material will determine how long it takes for it to travel through the material.

close-packed A crystal structure in which all of the atoms are as close together as possible.

colligative property Any property of a solution that depends on the concentration.

colloid Stable materials in which one type of particle is suspended in another without actually having been dissolved.

combustion When organic molecules combine with oxygen to form carbon dioxide, water vapor, and a large quantity of heat.

common ion effect When the addition of an ion affects the solubility or reactivity of a chemical compound.

compound Pure substances made up of two or more elements in defined proportions.

condensation The process by which a gas becomes a liquid.

conductor A material through which electricity can flow.

conjugate acid The compound formed when a Brønsted-Lowry base accepts a proton.

conjugate base The compound formed when a Brønsted-Lowry acid gives up a proton.

continuous spectrum Created when white light is broken up into a multicolored rainbow of light without gaps.

conversion factor A number that allows you to convert from one unit to another.

covalent bond Bonds created when two valence electrons are shared.

covalent compound Compound created when two or more atoms are held together with covalent bonds.

critical point The conditions of pressure ("critical pressure") and temperature ("critical temperature") past which the gas and liquid phases of a material can no longer be distinguished from one another.

crystal Large arrangements of ions or atoms that are stacked in regular patterns.

cycloalkane An alkane in which the carbon atoms are arranged in a ring.

d-transition metals The metallic elements in groups 3–12 of the periodic table.

decomposition reaction When large molecules break apart to form smaller molecules.

deposition The process by which a gas becomes a solid without first becoming a liquid.

determinate error *See* systematic error.

differential rate law A rate law that explains the relationship between the concentration of the reactants and the reaction rate.

diffusion The rate at which a gas travels across a room.

dilution The process by which a solvent is added to a solution to make the solution less concentrated.

dipole-dipole force An attractive force caused when the partially negative side of one polar molecule interacts with the partially positive side of another.

dissociation Fancy word for "dissolving."

distillation A process in which a mixture of materials is heated to separate them. One material will vaporize more quickly than the other, allowing them to be separated.

doping A method by which the conductivity of semiconductors is increased by adding a small amount of another element.

double displacement reaction A reaction that occurs when the cations of two ionic compounds switch places.

effusion The rate at which a gas escapes through a small hole in a container.

electrode The location of oxidation or reduction in a volatic cell.

electrolysis The process by which a current is forced through a cell in order to make a nonspontaneous electrochemical change occur.

electrolyte A compound that, when dissolved, causes water to conduct electricity.

electron Negatively charged particles that are found in the orbitals outside the nucleus of an atom.

electron affinity The energy change that occurs when a gaseous atom picks up an extra electron.

electron capture When an inner shell electron is captured by the nucleus, decreasing the atomic number by one.

electron configuration A list of orbitals that contain the electrons in an atom.

electronegativity A measure of how much an atom will tend to pull electrons away from other atoms to which it has bonded.

element A substance that cannot be chemically decomposed into simpler substances.

elementary reaction One of the steps in a reaction mechanism.

emission spectrum The pattern of light given off by an atom when energy is added to it.

endothermic A reaction that requires energy to occur.

endpoint The point where you stop a titration, generally because an indicator has changed color.

energy The capacity of an object to do work or produce heat.

energy diagram A graph that shows the amount of energy that the reactants have at all points throughout the chemical reaction.

enthalpy (H) The amount of heat present in a system at constant pressure.

entropy (S) A measure of the randomness of a system.

equation A shorthand way of describing a chemical process.

equilibrium When the concentrations of the products and reactants of a chemical reaction have stabilized because the rates of the forward and backward processes are the same.

equilibrium constant (K_{eq}) A constant that indicates whether the equilibrium will lie toward products or reactants.

equivalence point The point in a titration where $[H^+] = [OH^-]$.

excess reactant In a limiting reactant problem, the reactant that is left over when the limiting reactant has been completely used up.

excited state Any orbital with higher energy than the ground state.

exothermic A reaction that releases heat.

extraction A process by which a mixture of materials is shaken with a solvent to separate them. The separation occurs because one material will be more soluble in the new solvent than the other.

f-transition metals Another term for the lanthanides (elements 57–70) and actinides (elements 89–102).

family *See* group.

first law of thermodynamics *See* law of conservation of energy.

fission When an atomic nucleus breaks apart to make two smaller ones and a huge amount of energy.

free energy (G) Gibbs free energy, which is comprised of enthalpy (heat) and entropy (randomness). G is the fundamental measure that determines the position of equilibria and the rates of reactions. It is usually expressed in kJ/mol.

free radical An atom or group of atoms with an unpaired electron.

fusion A nuclear process in which small nuclei combine to make larger ones plus a huge quantity of energy.

gamma ray Very high energy electromagnetic radiation that's frequently given off when a nucleus undergoes radioactive decay.

gas The phase of matter in which particles are usually very far apart from one another, move very quickly, and aren't particularly attracted to one another.

geometric isomers Two or more structures that have the same formula and bond types (single, double, etc.) but differ in geometry (groups bonded to opposite sides of cyclic structures or double bonds). Geometric isomers are sometimes included in the generic term "stereoisomers."

ground state The orbital in which an electron is found if energy is not added to the atom.

group A column in the periodic table. Elements in the same group have similar chemical and physical properties.

half-cell The chemical process that takes place at one of the electrodes in a voltaic cell.

half-life ($t_{1/2}$) The amount of time it takes for half of the reactant to be converted to product in a first-order chemical or nuclear process.

half-reaction A reaction that shows only the oxidation or reduction process in a redox reaction.

halogens Elements in group 17 of the periodic table. These elements are the most reactive nonmetals.

heat The amount of energy that is transferred from one object to another during some process.

heat capacity *See* specific heat.

heterogeneous equilibrium An equilibrium in which the components are in different phases.

heterogeneous mixture A mixture in which the components are unevenly mixed.

homogeneous equilibrium An equilibrium in which all components are in the same phase.

homogeneous mixture A mixture created when two or more substances are so completely mixed with one another that it has uniform composition.

Hund's rule Electrons will stay unpaired whenever possible in orbitals with equal energies.

hybrid orbital An orbital formed by mixing two or more of the outermost orbitals in an atom together.

hydrate A compound (often ionic) to which water has been added.

hydrocarbon A molecule that contains only carbon and hydrogen.

hydrogen bond The attraction between a hydrogen atom that's bonded to nitrogen, oxygen, or fluorine and the lone pair electrons on the nitrogen, oxygen, or fluorine atom of a neighboring molecule.

hydrogenation reaction The reduction by addition of hydrogen to an unsaturated material.

hypothesis An educated guess about how a problem may be solved.

ideal gas A gas that follows all of the postulates of the kinetic molecular theory.

indeterminate error *See* random error.

indicator A compound used to indicate whether a solution is acidic or basic. Litmus (red=acid, blue=base) and phenolphthalein (clear=acid, pink=base) are two of the most commonly used indicators.

insulator A material through which electricity can't flow.

integrated rate law A rate law that describes how the concentrations of the reactants in a chemical reaction vary over time.

intermediate A chemical that was formed by one step in a reaction mechanism that will be consumed in another.

intermolecular force A force that holds covalent molecules to one another.

ion A particle with either positive or negative charge.

ion product constant (K_w) Equal to 10^{-14}, it's the product of the H^+ and OH^- concentrations in an aqueous solution.

ionic compound A compound formed when a cation and anion combine with one another.

ionization energy The amount of energy required to pull one electron off of an atom.

isomers Different molecules with the same formulas.

isotopes Atoms of the same element that have different masses. These different masses are due to differing numbers of neutrons in the nucleus.

K_a The acid dissociation constant, which describes the position of the equilibrium $HA \rightleftharpoons H^+ + A^-$.

kinetic energy Energy caused by the motion of an object.

kinetics The study of reaction rates.

law of conservation of energy Energy can neither be created nor destroyed in any process.

law of conservation of mass The weights of reactants in a chemical reaction are the same as the weights of the products. No matter what chemical changes may occur, matter is neither created nor destroyed.

law of definite composition A chemical compound contains the same elements in the same proportions by mass, regardless of how it was made.

law of multiple proportions When two elements form more than one chemical compound, the ratios of the mass of one element that combines with a fixed mass of the other element can be expressed as a ratio of small, whole numbers.

Le Châtelier's Principle If you change the conditions of an equilibrium, the equilibrium will shift in a way that minimizes the effects of whatever it is you did.

Lewis acid A compound that can accept electron pairs from another compound.

Lewis base A compound that can donate electron pairs to another compound.

Lewis structure A picture that shows all of the valence electrons and atoms in a covalently bonded molecule.

limiting reactant The reactant that runs out first in a chemical reaction, limiting the amount of product that can be formed.

line spectrum When only certain bands of light are emitted in a spectrum.

liquid The form of matter in which molecules move around freely but still experience attractive forces.

London dispersion forces Temporary dipole-dipole forces created when one molecule with a temporary dipole induces another to become temporarily polar.

lone pair A pair of electrons in a compound that aren't involved in bonding.

magnetic quantum number Denoted by m_l, it determines the direction that the orbital points in space. Possible values for m_l are all the integers from $-l$ through l.

mass A measure of how much matter is present in an object. Mass is usually measured in grams.

mass defect The nucleus of an atom weighs less than the sum of the weights of the protons and neutrons. This is because some of this mass (called the "mass defect") has been converted to nuclear binding energy.

mass spectrometry The modern process by which the molecular formulas of new compounds are determined.

mechanism The process through which reactants form products.

metal A material that's shiny, malleable, ductile, and conducts electricity well.

molality (m) Moles of solute per kilograms of solvent.

molar mass The weight of 6.02×10^{23} atoms or molecules of a compound in grams.

molar volume The volume of one mole of any gas at standard temperature and pressure.

molarity (M) Moles of solute per liters of solution.

mole 6.02×10^{23} things.

mole fraction (χ) The number of moles of one component in a solution divided by the total number of moles of all components in the mixture.

mole ratio The ratio of moles of product to the ratio of moles of reactant of a chemical reaction.

molecular solid A material consisting of many covalent molecules held together by intermolecular forces.

molecular weight *See* molar mass.

molecularity The number of reactant molecules that combine in a chemical process.

molecule A group of atoms held together with covalent bonds.

network atomic solid A material in which many atoms are bonded together covalently to form one gigantic molecule.

neutrons Neutral particles with a mass of about 1 amu that are found in the nucleus of an atom.

noble gases Group 18 on the periodic table, noted for unreactivity.

normal boiling point The temperature at which a liquid boils at a pressure of 1.00 atm.

normality (N) The number of moles of a reactive species per liter of solution.

nucleon The particles in the nucleus of an atom, namely protons and neutrons.

nucleus The center of an atom where the protons and neutrons are found.

nuclide A word used to describe a particular isotope of an element.

octet rule Elements tend to want to gain or lose electrons to attain the same electron configurations as the nearest noble gas.

orbital Regions of space outside the nucleus of an atom in which electrons can be found.

order An exponential term in a rate law that describes how the overall rate of the reaction depends on the concentration of each reactant.

organic compound Covalent compound that contains carbon.

oxidation state Also called the "oxidation number," this is the charge that the atom is considered to have in a chemical compound.

oxidize To lose electrons.

oxidizing agent A compound that causes another to be oxidized—in the process, it is itself reduced.

partial pressure The partial pressure of one gas in a mixture of gases is equal to the amount of pressure that would be exerted by that gas alone if all of the other gases were removed.

parts per million (**ppm**) The number of mg (0.001 g) of solute by the number of liters of water.

Pascal (**Pa**) The metric unit of pressure. There are 1.01325×10^5 Pa in 1 atm.

Pauli exclusion principle No two electrons in an atom can have the same four quantum numbers.

period A horizontal row in the periodic table. Elements in the same period have valence electrons with similar energies.

pH The scale used to indicate the acidity of a solution, defined as $-\log[H^+]$.

phase The state of matter, either solid, liquid, or gas. Solids don't mix at all, liquids can form several immiscible phases, and all gases mix to form one phase.

phase diagram A graph that shows in what phase a material can be found at all combinations of temperature and pressure.

pKa $-\log(K_a)$.

plum pudding model An early model of the atom in which small bits of negative charge (electrons) are embedded in a giant blob of positive charge.

polar A term referring to a molecule that has partial positive charge on one side and partial negative charge on the other.

polar covalent bond A covalent bond in which the electrons aren't shared equally between both atoms.

polyatomic ion An ion containing more than one atom.

polymerization When small molecules called monomers link up to form a very long molecule called a polymer.

positron The antimatter equivalent of an electron. It has a positive charge and is created during positron decay.

potential energy Stored energy. In chemical processes, it's frequently stored in chemical bonds.

precision When a value can be measured repeatedly. High precision in a measurement usually (but not always) indicates high accuracy.

pressure The amount of force exerted by the particles in a gas as they hit the sides of a container.

principal quantum number Denoted by "n," it describes the energy level of an electron. Possible values are 1, 2, 3… n.

probability distribution A graph that shows how the electron density of an orbital is related to the distance from the nucleus.

product The final result of a chemical reaction.

protons Positively charged particles with a mass of about 1 amu that are found in the nucleus of an atom.

qualitative data Observations that can't be expressed as numbers ("My cat is ugly").

quantitative data Any measurements that involve numbers, such as weights, lengths, times, or temperatures.

radiation The small particles emitted during the radioactive decay of an atomic nucleus.

radioactive decay When a nucleus spontaneously breaks apart to form smaller particles.

radioisotope Any radioactive isotope.

random error Unpredictable sources of error that can't be compensated for.

rate constant (k) A constant, unique for every chemical reaction, which indicates how quickly it will form products from reactants.

rate-determining-step The elementary step in a reaction mechanism that proceeds most slowly.

rate law An expression that shows how the rate of a chemical reaction depends on the concentration or temperature of the reactants.

reactant The starting ingredient for a chemical reaction.

reaction order The sums of the orders of all reactants in a chemical reaction.

redox reaction A reaction in which the oxidation state of the reactants changes.

reduction The process of gaining electrons.

reduction agent A compound that causes another to be reduced—in the process, it is itself oxidized.

resonance structures Lewis structures in which the positions of the electrons or bonds in a molecule are changed, but the atoms remain in the same locations. Resonance structures are convenient, though imaginary, ways of representing the true form of the molecule, known as the resonance hybrid.

reversible reaction A reaction in which the reactants form products and the products reform reactants.

root-mean-square (RMS) velocity The average velocity of the molecules in a gas.

salt Generic term for an ionic compound.

salt bridge A tube containing an ionic compound that allows charge transfer in a voltaic cell.

saturated hydrocarbon Fancy way of saying "alkane."

saturated solution A solution that has dissolved the maximum possible amount of solute.

scientific method A systematic method of solving problems based on experiments and observations.

second law of thermodynamics The entropy change is always positive for spontaneous processes.

semiconductor A material through which electricity only flows well at high temperatures or voltages.

SI units The standard metric system of units.

significant figures The number of digits in a measured or calculated value that gives us meaningful information about the thing being measured or calculated.

single displacement reaction When a pure element switches places with one of the elements in a chemical compound.

solid The state of matter in which the atoms or molecules are locked into place by either chemical bonds or intermolecular forces.

solubility product constant (K_{sp}) The equilibrium constant for the dissociation of a solute into a solvent.

solute The thing that gets dissolved in a solution.

solution *See* homogeneous mixture.

solvent The major component that dissolves a solute. Solvents are usually liquids, but can also be solids.

specific heat (C_p) The amount of energy required to heat one gram of a substance by one Kelvin at constant pressure.

spectroscopy A method of identifying unknown substances from their spectra.

spectrum A pattern of light that corresponds to the movement of electrons between the ground and excited state. The plural of spectrum is "spectra."

spin quantum number Denoted by "m_s," it distinguishes between the two electrons in an orbital. Possible values are $+\frac{1}{2}$ or $-\frac{1}{2}$.

spontaneous A process that takes place without outside intervention.

standard conditions For gases, 1 atm and 273 K; for liquids, 1 M and 273 K.

standard temperature and pressure (STP) 0° C (273 K) and 1 atm.

stereoisomers Isomers that differ in three-dimensional structure from one another.

stoichiometry The method we use to relate the masses or volumes of the reactants and products of a chemical reaction to each other.

strong When used to describe acids, bases, or electrolytes, this adjective means that the compound in question is fully dissociated into component ions.

structural isomers Compounds having the same formula but differing in functional group or bonding pattern.

sublimation The process by which a solid becomes a gas without first becoming a liquid.

supercritical fluid A material at high enough conditions of temperature and pressure that it's no longer clear whether it's a gas or liquid.

supersaturated solution A solution that has dissolved more than the normal maximum possible amount of solute.

surface tension The tendency of liquids to keep a low surface area.

synthesis reaction When small molecules combine to form larger ones.

systematic error An error that causes experimental data to be skewed by the same amount every time.

temperature A measure of the quantity of kinetic energy present in an object.

thermodynamics The study of free energy, enthalpy, and entropy.

titration The use of neutralization reactions to determine the concentration of an acid or base.

transition state The highest energy state between products and reactants in a chemical reaction.

triple point The conditions of temperature and pressure at which the liquid, gas, and solid phases of a material are all stable.

unit cell The smallest unit that can be stacked together to re-create a crystal.

unsaturated hydrocarbon A hydrocarbon containing at least one multiple bond.

unsaturated solution A solution that hasn't yet dissolved the maximum possible quantity of solute.

unshared electron pair *See* lone pair.

valence electrons The number of s- and p- electrons beyond the most recent noble gas.

Valence Shell Electron Pair Repulsion theory (VSEPR) The shapes of covalent molecules depend on the fact that pairs of valence electrons tend to repel each other.

vapor pressure The vapor pressure of a liquid is the gas pressure in a closed container due to the molecules that have evaporated from the liquid.

voltaic cell Fancy word for "battery."

volume A measure of how much space an object occupies. In chemistry, it's usually measured in milliliters or liters.

weak Used to describe acids, bases, and electrolytes, it indicates that the compound in question only partially dissociates in water.

Index

C

D

E

F

G

H

I-J

K

L

M

N

O

P

Q

R

S

T

U

V

W

X-Y-Z